1st International Symposium on Spintronic Devices and Commercialization 2010

(ISSDC 2010)

AA002796

Journal of Physics: Conference Series Volume 263

Beijing, China
21 – 24 October 2010

ISBN: 978-1-61782-298-8
ISSN: 1742-6588

Printed from e-media with permission by:

Curran Associates, Inc.
57 Morehouse Lane
Red Hook, NY 12571

Some format issues inherent in the e-media version may also appear in this print version.

Printed by Curran Associates, Inc. (2011)

For permission requests, please contact the Institute of Physics
at the address below.

Institute of Physics
Dirac House, Temple Back
Bristol BS1 6BE UK

Phone: 44 1 17 929 7481
Fax: 44 1 17 920 0979

techtracking@iop.org

Additional copies of this publication are available from:

Curran Associates, Inc.
57 Morehouse Lane
Red Hook, NY 12571 USA
Phone: 845-758-0400
Fax: 845-758-2634
Email: curran@proceedings.com
Web: www.proceedings.com

TABLE OF CONTENTS

PREFACE

ISSDC'2010

国际磁电子器件及产业化研讨会

INTERNATIONAL SYMPOSIUM ON SPINTRONIC DEVICES AND COMMERCIALIZATION

Preface

The International Symposium on Spintronic Devices and Commercialization (ISSDC' 2010) was held in Beijing, China, from 21 to 24 October 2010. The aim of the symposium was to provide an opportunity for international experts, academics, researchers, practitioners and students working in the areas of spintronic theories, spintronic materials, and spintronic devices to exchange information on the R&D and commercialization of spintronic materials and devices. New developments, concepts, future research trends and potential commercialization areas were also discussed. The topics covered by ISSDC' 2010 were:

- **Fundmental Spintronic Theories/Experiments**
  Spin polarization, spin-dependent scattering, spin relaxation, spin manipulation and optimization, as well as other related characterizations and applications, etc.
- **Spintronic Materials**
  Giant magnetoresistance materials, magnetic tunnel junction materials, magnetic semiconductor materials, molecular spintronic materials.
- **Spintronic Devices**
  Sensors, isolators, spin logic devices and magnetic random access memories (MRAMs), microwave devices, spin diodes, spin transistor, spin filters and detectors, spin optoelectronic devices, spin quantum devices, single chip computer, spin molecule and single electron devices.
- **Other Magnetic Materials**
  Soft magnetic materials, hard magnetic materials, magneto-optical materials, magnetostriction materials.
- **Applications of Spintronic Devices**
  Magnetic position/angle/velocity/rotation velocity sensors, magnetic encoders, magnetic compasses, bio-medical magnetic devices and other applications.
- **Future Research Trends and the Commercialization of Spintronic Devices**

Approximately 85 scientists from almost 10 countries participated in the conference. The conference featured 6 keynote lectures, 8 invited lectures, 12 contributed lectures and about 30 posters.

We would like to express our gratitude to all participants for their presentations and discussions, which made the conference very successful indeed. We are also grateful to the Advisory Committee, the Conference Chairs, the Excutive Chairs, the Academic Committee, the Organization Committee and the Secretariat for their fruitful work. We would especially like to thank all the organizers listed below for their support in all aspects of the conference.

We would like to express our thanks to all the authors for their time and genuine efforts, and to the reviewers for their fruitful comments during the preparation of this volume.

ISSDC'2010 ORGANIZATION

		USA	
	Jungchun Huang, Taiwan	Dexuan Huo, China	Yoon H Jeong, Korea
	Chihuang Lai, Taiwan	Runwei Li, China	Wei Liu, China
	Jing Shi, USA	Yasushi Takemura, Japan	Mark Tondra, USA
	Shan X Wang, USA	Di Wu, China	Desheng Xue, China
	Minglang Yan, USA	Shishen Yan, China	Xiaofei Yang, China
	Chun Yeol You, Korea	Wei Zhao, China	Shiming Zhou, China
	Jianguo Zhu, China		

Secretariat

Secretary-general	Changmao Yang, China		
Vice Secretary-general	Junli Wang, China	Jinsong Xu, China	Ye Tian, China
Members	Ru Bai, China	Hongliang Zhan, China	

ISSDC' 2010 Organizers

Department of Science and Technology, CSIC, China

SpinIC Inc., China

Hangzhou Dianzi University, China

State Key Laboratory of Magnetism, Institute of Physics, Chinese Academy of Sciences, China

Institute of Measurement Technology, China

Magnetic Materials and Devices Branch of Electronic Component Association, China

Shenyang Academy of Instrumentation Science / National Engineering Research Center for Transducer, China

Journal of Physics: Conference Series **263** (2011) 012001 doi:10.1088/1742-6596/263/1/012001

Resonant circuit as magnetic device for cancer therapy

Yasushi Takemura

Department of Electrical and Computer Engineering, Yokohama National University
79-5 Tokiwadai, Hodogaya, Yokohama, 240-8501, Japan

E-mail: takemura@ynu.ac.jp

Abstract. Fabrication and characterization of a resonant circuit as magnetic device for hyperthermia treatment are discussed. The resonant circuit consisting of a closed connection of an inductor and a capacitor raised its temperature by an externally applied magnetic field. As the resonant circuit was heated efficiently, it could be excited by a weak RF magnetic field of magnetic resonance imaging (MRI). By using a ferrite core for the inductor, the efficiency of temperature rise of the circuit was improved, which attributed to the miniaturization of the implant size.

1. Introduction

Hyperthermia is a therapeutic procedure of raising the body temperature for cancer treatment. Various established treatments of surgical operation, radiotherapy and chemotherapy suffer from risks of scar and negative side effects. The hyperthermia is expected for reducing those risks. The hyperthermia elevates the local temperature above 42.5 deg C. to kill cancer cells, while most normal tissues are not damaged at the temperature below 44-45 deg. C [1]. When the body temperature rises, the heat is dissipated by cooling through blood circulation. A sluggish blood flow in the cancerous tissues leaves them vulnerable at the elevated temperature. This process of killing the cancer is safe for surrounding healthy tissues.

A technique for local heating is essential in terms of capability of warming the cancerous tissue locally as compared with whole body heating. Among various warming methods proposed for the local heating, e.g. magnetic nanoparticles [2], the hyperthermia using implants is expected to warm cancerous tissue in deeply located. The implants of tiny elements inside of the body are heated up by an external energy source. A combination of magnetic materials as the implants and a magnetic field applied from outside of the body has attracted much attention [3, 4]. In order to minimize harmful effects associated with high frequency magnetic field to the body, the external magnetic field should be less intense with a lower frequency as possible. It is expectable to use a resonance circuit for the implant, which generates heat efficiently by the applied external magnetic field [5]. It was found that the resonant circuit was heated up to 40 deg. C by the weak applied magnetic field of 3.4 μT at 63.9 MHz [6]. It has been also proposed to use a RF pulse (high frequency magnetic field) from a magnetic resonance imaging (MRI). As a preliminary result, RF magnetic field of the MRI with a low duty factor could raise the circuit temperature up to 13 deg. C [7].

In this paper, the resonant circuit as magnetic device for cancer therapy is discussed in detail, including excitation by RF field from the MRI, higher efficiency of heating using magnetic core in inductors, and miniaturization for delivering through catheter or injector as lower invasive process.

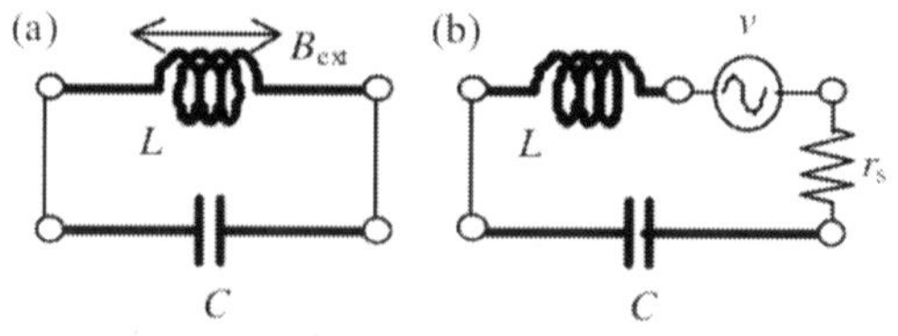

Figure 1. Resonant circuit consisting of closed connection of inductor and capacitor (a) and its equivalent circuit under ac magnetic field (b).

2. Resonant circuit as hyperthermia implant

The resonant circuit used in this study was consisted of a closed connection of an inductor and a capacitor. When an ac magnetic field is applied to the inductor coil, a current flows by an electromotive force induced in the coil, v as shown in Fig. 1. An electric power consumed in the circuit, P in active power, is described as

$$P = \frac{r_s}{Z^2} v^2 \qquad (1)$$

where Z and r_s are impedance and residual resistance of the circuit, respectively. In case that the frequency of the applied ac field equals to the resonant frequency of the circuit, the power P is described as

$$P_{resonance} = \frac{1}{r_s} v^2 \qquad (2)$$

because the reactance in $Z(= j\omega L + \frac{1}{j\omega C} + r_s)$ is zero.

As the electromotive force induced in the coil, v is proportional to the frequency of the applied magnetic field, higher frequency is preferable for a higher temperature rise.

The circuit was consisted of an inductor (3.4 μH) and a capacitor (18 pF). Its resonant frequency was 63.9 MHz. The diameter and number of turn of the coil were 6 mm and 9, respectively. The temperature rise of the circuit was measured by an optical thermometer, whose result was not affected by a magnetic field with high frequency. Its sensor tip was attached to the coil. Preliminary results of the temperature rise of the circuit excited by an ac magnetic field with continuous waveform have been reported [6]. It was found that the highest temperature rise was obtained when the frequency of the applied field matched to the resonant frequency of the circuit.

3. Temperature rise of resonant circuit

3.1. Excitation by MRI

A temperature rise of a resonant circuit excited by a RF pulse of MRI was measured. The 1.5-T type commercial MRI system was used. The center frequency of the modulated RF magnetic field was 63.9 MHz. The magnetic field of the modulated RF pulse was applied to the circuit under a normal procedure for the MRI diagnostics. The direction of the inductor coil was adjusted to the x-, y- or z-axis, where the z-axis was parallel to the direction of the dc magnetic field in the configuration of the MRI system. The RF pulse was applied in the directions of the x- and y-axes which were perpendicular to the z-axis. Figure 2 shows the temperature rise of the circuit as a function of the duration of the MRI operation. The circuit raised its temperature up to 20-30 deg C. after 5 min when the coil was aligned to x- or y-axis which was parallel to the direction of the RF pulse. Field strength of RF pulse of MRI for clinical use in hospitals is normally a few μT and its duty ratio is quite low comparing with a continuous wave. It should be noted that the resonant circuit is heated by this weak RF magnetic field. The temperature did not rise in case that the RF magnetic field did not penetrate the coil aligned to z-axis. The slight difference of temperature rise between x- and y-axes were due to different power cycle

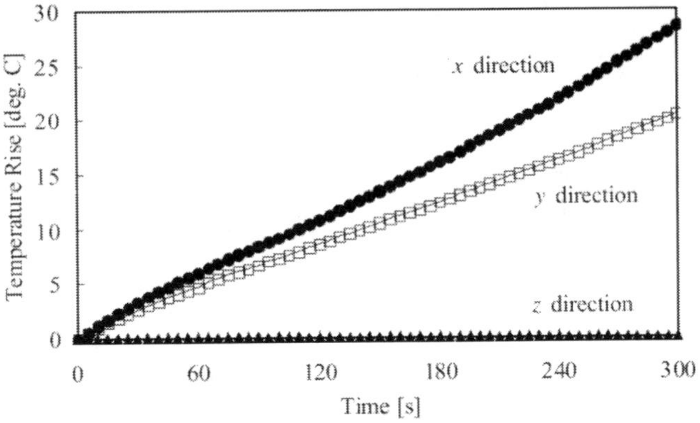

Figure 2. Temperature rise of resonant circuit excited by RF pulse of MRI. The inductor was aligned to various directions in MRI.

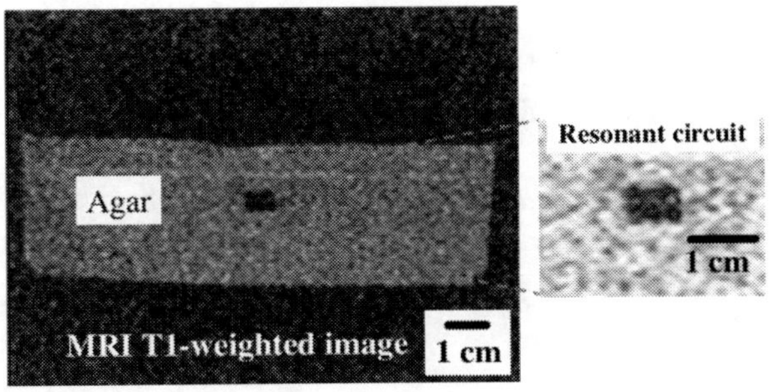

Figure 3. T1-weighted MRI image of agar phantom with inserted resonant circuit implant.

in MRI diagnosis sequence.

Because the inductor acts as a magnet, the resonant circuit implant might degrades the MRI images as artifacts. Figure 3 shows the T1-weightened MRI image of an agar phantom with an inserted resonant circuit. The agar was semi-hard, but the position of the circuit was not shifted in the MRI. Although the part of the tomographic image around the circuit was blackened, the surrounding part was clearly imaged enough for diagnosis.

3.2. Frequency modulation in MRI

The waveform of RF magnetic field is modulated sinc wave pulse in order to observe recovery and decay of hydrogen nuclei magnetic moment. Its frequency is distributed as to match the nuclear magnetic resonance (NMR) frequency, also called the Larmor frequency, for each position. Figure 4 shows the waveform of RF pulse and its frequency distribution. They are related to each other in Fourier and Fourier inverse transformations. In order to collect tomographic images, information depending on the position under gradient dc magnetic field corresponds to information depending on NMR frequency. In MRI equipment used in this study, the frequency shift of 0.1 MHz can be obtained by the position shift of 70 mm by employing an appropriate sequence. The temperature rise of the resonant circuit depending on the position in the MRI stage was investigated as shown in Fig. 5. The resonant circuit was placed at distance of 0, ±70 and ±140 mm from the center position and its temperature rise of measured at each position. The frequency at the center was 63.9 MHz. Figure 6 shows the temperature rise measured at 30 sec and 300 sec after irradiating the RF pulse. It was confirmed that temperature rise was maximum at the center position, where the excitation frequency was matched to the resonant frequency of the circuit of 63.9 MHz. From this result, it is suggested that

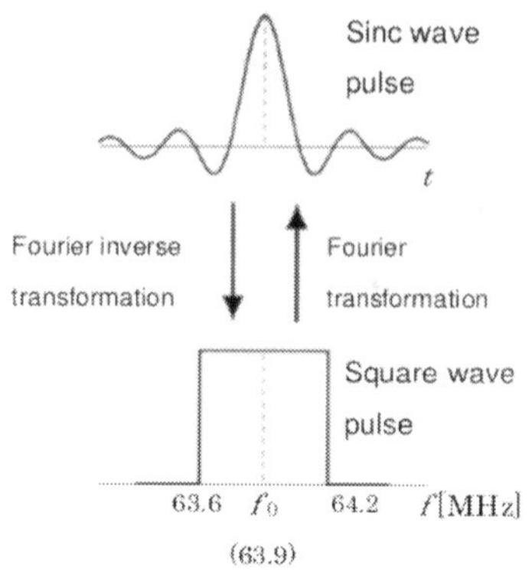

Figure 4. Waveform of RF magnetic field as sinc wave pulse and its frequency distribution.

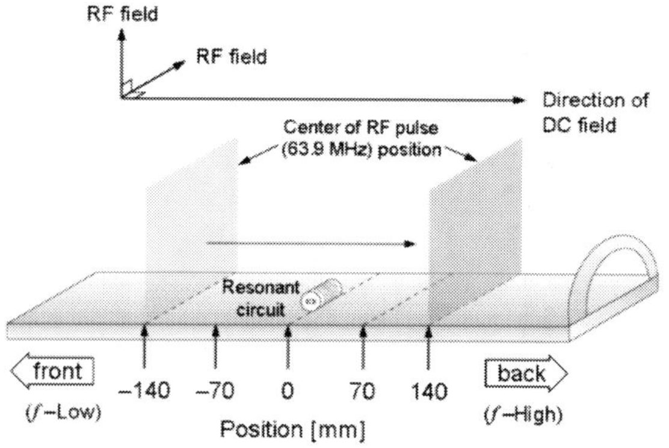

Figure 5. Frequency of RF pulse depending on the irradiation position of MRI equipment.

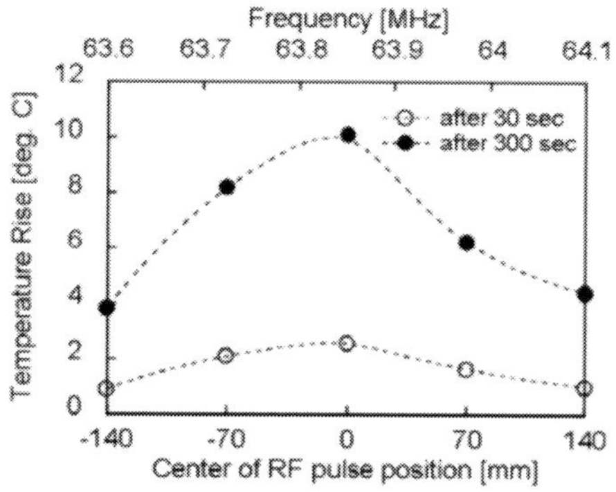

Figure 6. Frequency dependence of temperature rise of the resonant circuit measured by changing its position.

the temperature of the resonant circuit implant can be controlled by its position as well as by pulse duty ratio of MRI sequence.

3.3. Resonant circuit using core-coil

In case that size of implants is as small as 1mm, it is possible to deliver them to a tumor through a catheter as illustrated in Fig.7. This delivery method offers a cancer treatment with less scar, which can not be achieved by surgical operation including endoscopic resection. A risk of infectious disease can be also reduced. A resonant circuit is one of the promising implant from the viewpoints of its size and high heat efficiency.

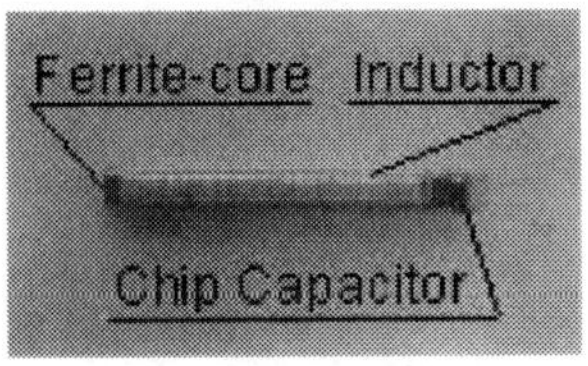

Figure 7. Resonant circuit using core-coil. The diameter of this implant was reduced to less than 1 mm.

Recently miniaturization of the resonant circuit implant has been achieved as shown in Fig. 7. Previously, the inductor was air-core type. The Ni-Zn ferrite core of a cylindrical shape was employed for the inductor in order to enhance the induced electromotive force from the external ac magnetic field. As a preliminary result, it was found that the diameter of the resonant circuit implant using a core-coil could be reduced to a tenth of that using an air-coil. The reduced size less than 1mm in diameter allows delivery of the implant into the body using an injector as a lower invasive process.

4. Conclusions

The cancer therapy of hyperthermia using a resonant circuit implant us discussed. The implant consisting of a closed connection of an inductor and a capacitor was efficiently heated by an external ac magnetic field. It was found that normal MRI equipment of clinically used in hospital system could rise the temperature of the resonant circuit. The temperature rise was up to 30 deg. C by an applied magnetic field of RF pulse under a normal diagnosis procedure of the MRI system. It is advantageous that the commercial MRI equipment can be used for the hyperthermia because the therapeutic procedure can be repeated after confirming results from the MRI diagnosis. The size of the implant could be reduced by using the core-coil. This recent improvement is expected for delivery of the implant through a catheter or an injector.

References

[1] W. C. Dewey, L. E. Hopwood, S. A. Sapareto, L. E. Gerweck, *Radiology* **123**, 463-474 (1977).
[2] Q. A. Pankhurst, J. Connolly, S. K. Jones and J. Dobson, *J. Phys. D: Appl. Phys.*, **36**, R167-R181 (2003).
[3] I. Tohnai, Y. Goto, Y. Hayashi, M. Ueda, T. Kobayashi and M. Matsui, *Int. J. Hyperthermia*, **12**, 37-47 (1996).
[4] M. Jojo, A. Murakami, F. Sato, H. Matsuki, and T. Sato, *IEEE Trans. Magn.*, **37**, 2944-2946 (2001).
[5] Y. Kotsuka, K. Orii, H. Kojima, K. Kamogawa, and M. Tanaka, *IEEE Trans. MTT*, **47**, 2630-2635 (1999).
[6] M. Morita, T. Inoue, T. Yamada, Y. Takemura, T. Niwa, and T. Inoue, *IEEE Trans. Magn.*, **41**, 3637-3675 (2005).
[7] T. Niwa, Y. Takemura, T. Inoue, H. Kurihara, T. Hisa, *British J. of Radiology*, **81**, 69-72 (2008).

Journal of Physics: Conference Series **263** (2011) 012002 doi:10.1088/1742-6596/263/1/012002

Detection of magnetic microbeads and ferrofluid with giant magnetoresistance sensors

J Feng, Y Q Wang, F Q Li, H P Shi and X Chen

National Key Laboratory of Micro/ Nano Fabrication Technology, Key Laboratory for Thin Film and Microfabrication Technology of Ministry of Education, Institute of Micro and Nano Science and Technology, Shanghai Jiaotong University，Shanghai 200240 PR China

E-mail: jfeng@sjtu.edu.cn

Abstract. Giant magnetoresistance sensors based on multilayers $[Cu/NiFeCo]_{\times 10}$/ Ta were fabricated by microfabrication technology. A GMR-bridge was used to detect the magnetic MyOne beads and Ferro fluid. The dependence of the GMR-bridge signals on the surface coverage of MyOne beads was studied. The results show that the GMR sensor is capable of detecting the magnetic beads. The detectable limit of MyOne beads is about 100, and the corresponding signal output is 8 μV. The GMR bridge signal is proportional to the surface coverage of the MyOne beads. The sensitivity of the GMR bridge is inversely proportional to the feature size of the GMR sensor. The GMR bridge integrated with microfludic channel was also used for dynamic detection of ferrofluid (suspension of Fe_3O_4 particles). The results show that the GMR bridge is capable of detecting the flow of ferrofluid, and the sensor signals are proportional to the concentration of the ferrofluid. The detection limit of concentration of the ferrofluid is 0.56 mg/ml, and the corresponding signal is 6.2 μV.

1. Introduction

Cheap, reliable and portable biosensing technologies for medical applications have received growing attention. Different techniques are employed to develop lab-on-a-chip system, such as fluorescent labels with an exciting laser and detection optics and electrochemical detections. In these techniques, micro-arrays can be integrated to detect the multi analytes simultaneously.

Recently magnetic particles have been developed as labels for biosensing, due to their several advantages over other labels [1,2]. Magnetoresistance devices including giant magnetoresistance (GMR) sensor [3-5], spin-valve GMR sensor [6-8], tunneling magnetoresistive sensor [9], and Hall Effect sensor [10] are widely studied and employed for detection of magnetic labels. The GMR sensors can be integrated into micro-arrays for multi-analyte detection. The GMR sensors can also be integrated with microfluidics to realize more complex bioassay [11]. By applying a magnetic field on chip, the bio-molecules bound with magnetic labels can be manipulated and separated on chip and finally detected by the integrated GMR sensors [12].

In this study, multilayer-based GMR sensors were fabricated. A GMR-bridge was constructed and employed to detect the magnetic MyOne beads. The GMR sensor was also integrated with microfludic channel for dynamic detection of ferrofluid. The dependence of the GMR-bridge signals on the surface coverage of MyOne beads and the concentration of ferrofluid were studied.

2. Experiment

Giant magnetoresistance (GMR) multilayers were deposited by dc magnetron sputtering onto 3 inch-Si wafers with the structure of NiFeCo 6nm[Cu 2.1nm/NiFeCo 1.5nm]$_{\times10}$/ Ta 100nm. The sensors based on the GMR multilayers were fabricated by lithography, ion beam etching and lift-off technology. The sensors were covered with a 200 nm thick sputtered SiO_2 passivation layer to protect it from environmental corrosion. Finally a layer of SU-8 photoresist was coated on the chip. A micro channel on the GMR sensor with 200 μm wide and 50 μm in depth was formed by photo lithography. The structure of a GMR sensor element is shown in figure 1(a), and the magnetoresistance response to an in-plane magnetic field of the sensor element is displayed in figure 1(b). The GMR sensors revealed magnetoresistance ratio of ~9.8%. The resistances of the sensor elements are ~5 kΩ and ~3 kΩ for the sensors with the feature line width of 3 μm and 5 μm respectively.

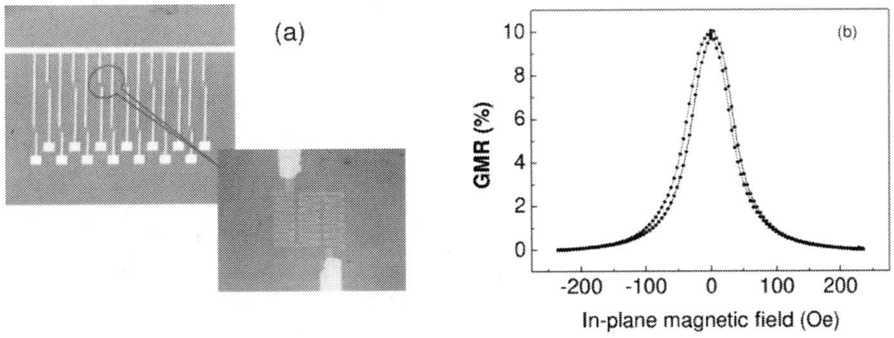

Figure 1. The structure of a GMR sensor element (a) and the magnetoresistance ratio of the GMR sensor (b).

A GMR-bridge, resembling a Wheatstone bridge, was constructed by connecting two GMR sensors and two external adjustable resistors for the detection of magnetic beads and ferrofluid. One of the two GMR sensors was used as sensing element and the other one was used as reference. The GMR-bridge was biased by 1-2 V DC current generated by 2400 Sorcemeter. The voltage signal across the bridge was measured directly using an Agilent multimeter and the acquired data fed to a computer. A vertical magnetic field of 240 Oe generated by a Helmholtz coil was applied as induced field in the detection of magnetic beads, while an in-plane magnetic field of 70 Oe was applied in ferrofluid detection. These magnetic fields were applied to induce a moment within the superparamagnetic particles. In dynamic detection of ferrofluid, a GMR sensor chip was embedded in a PDMS base at first, after the PDMS base solidified the micro channel was constructed by aligned bonding SU-8 layer and PDMS base to PDMS lid.

3. Results and discussion

3.1. Detection of magnetic beads

The superparamagnetic MyOne beads from Invitrogen Company with diameter of 1 μm were used and a DC magnetic field of 240 Oe normal to the sensor plane was applied in this section. Because of the demagnetizing effect, GMR sensors do not respond to the vertical magnetic field. When the sensing area of GMR bridge was free of magnetic beads, almost no resistance change in sensing element and no voltage output in GMR bridge was observed even if the normal magnetic field increased from 0 to 240 Oe. The calibration of GMR bridge was performed by adjusting the external resistor at first. When no magnetic beads are on the sensing area, the voltage output of the GMR bridge at this time is considered to be zero point of signal. Then some MyOne beads were settled on the surface of the sensing area, a vertical magnetic field of 240 Oe was applied. A signal output of the GMR bridge was observed owing to the occurrence of the MyOne beads, which means that the GMR bridge is capable of detecting the existence of the MyOne beads.

The magnetic liquid with various concentrations of MyOne beads was dropped and various content of MyOne beads were settled on the surface of the GMR sensors. The signal response of GMR-bridge to the MyOne beads that were on the surface of the sensing area was measured. Figure 2 is the dependence of the signal output on the surface coverage of MyOne beads. The surface coverage of MyOne beads was calculated by Photoshop soft ware. The GMR bridge signal is almost proportional to the surface coverage of the MyOne beads. In figure 2, deviations of the signal to the linearity can be observed. The deviations to the linearity can be attributed to several factors, such as the tilt of the induced magnetic field, the deviation of the position of the beads from the central area of the sensor strip, and the aggregation of MyOne beads as shown in figure 3.

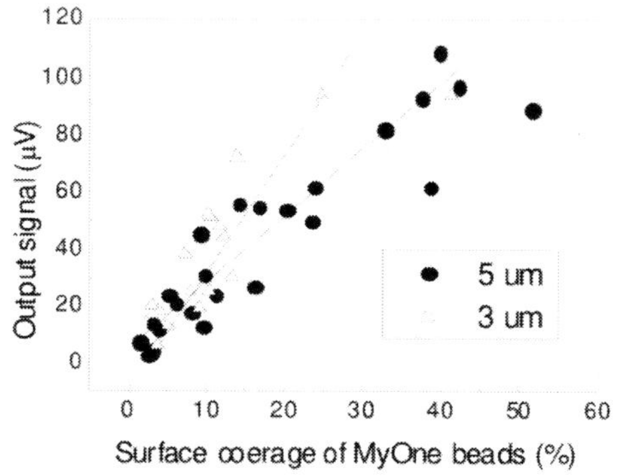

Figure 2. The dependence of the output signal on the surface coverage of MyOne beads.

The impact of feature size of the GMR sensor on its detection sensitivity was also studied, and the signals of the GMR sensors with line width of 5μm and 3 μm are all displayed in figure 2. The result shows that the GMR sensor with line width of 3 μm is more sensitive than that one with line width of 5 μm. The sensor sensitivity increases while the feature size of the GMR sensor decreases.

Figure 3 is the images of the MyOne beads on the surface of sensing area with different surface coverage and signal output. The more the beads covered the sensing area, the larger the signal output of GMR bridge was. The beads number was counted directly when the surface coverage was 1.37%, and the MyOne beads number was about 100, and the corresponding signal output of GMR bridge was 8 μV. In this study, the detection limit of the number of MyOne beads is about 100, which is in agreement with the reported results [5]. Target biomolecules such as DNA can be marked with MyOne beads by hybridization assays, therefore the GMR sensors can be applied in biosensor system to detect the biomolecules by measuring the density of magnetic beads.

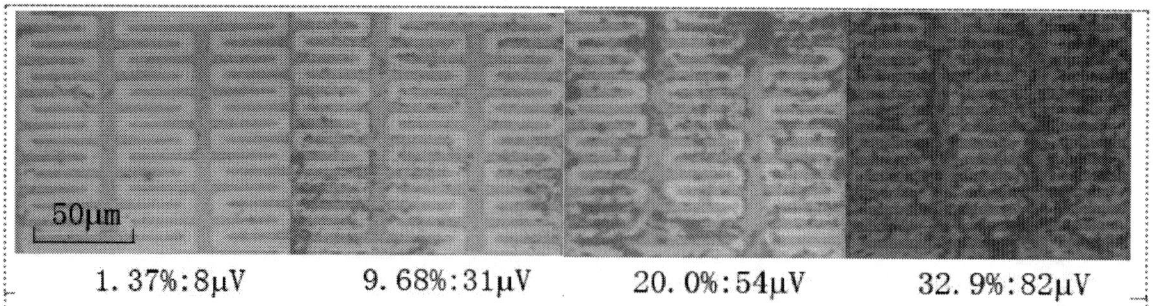

Figure 3. The images of the MyOne beads on the surface of sensing area and the output signals.

3.2. Dynamic detection of ferrofluid

In order to dynamically detect the magnetic particles in flow, the GMR sensors were integrated with a microfludic channel which was constructed by bonding SU-8 channel layer to PDMS lid. Figure 4 is sectional schematic illustration of a GMR sensor bound with a microfludic channel. The width of the microfludic channel is 200 μm and the depth is 50 μm. The GMR sensing element is directly under the microfludic channel, while the reference GMR sensor is away from the channel. Ferrofluid (suspension of ~10 nm Fe_3O_4 superparamagnetic particles) was used in this section.

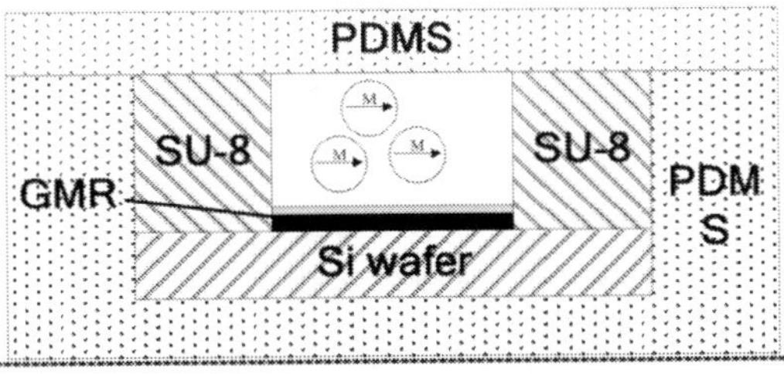

Figure 4. The schematic illustration of a GMR sensor bound with a microfluidic channel.

Syringe pump was connected to microfludic channel by capillaries via access ports that were formed in the PDMS lid. A ferrofluid plug and a water plug were pushed into micro channel and passed through the GMR sensing area alternately. The flow rate of ferrofluid was 5 μl /s. An in-plane magnetic field was applied to induce the stray field of magnetic particles. The signal of GMR bridge is proportional both to the sensitivity of the GMR sensor and to the stray fields of magnetic particles. Because an external field of 70 Oe can induce larger stray fields of magnetic particles than the magnetic field of 50 Oe, an in-plane magnetic field of 70 Oe was applied, even if the sensitivity of the GMR sensor is larger at the magnetic field of ~50 Oe. The dynamic detection was performed by directly detecting the stray fields generated by Fe_3O_4 particles in the ferrofluid plug while it passed through the sensing area. Real time GMR bridge signals were directly measured using a GPIB-controlled multimeter and the acquired data fed to a computer.

Figure 5 shows the real time response of the GMR bridge to the ferrofluid. When a ferrofluid plug passed through the sensing area, a voltage signal of ~40 μV was observed. The four signal peaks in figure 5 indicate the four ferrofluid plug passing through the GMR sensing area serially. We repeated the dynamic detection with ferrofluid of different concentration. Figure 6 shows the relationship of the GMR signal to the concentration of ferrofluid. The GMR bridge signals are proportional to the concentration of the ferrofluid.

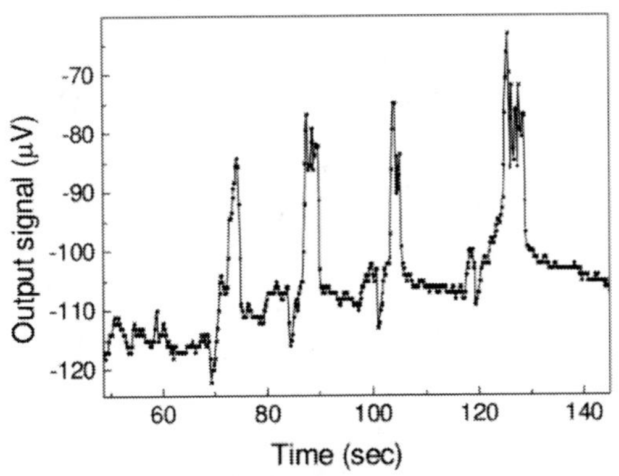

Figure 5. The real time response of the GMR bridge to the ferrofluid.

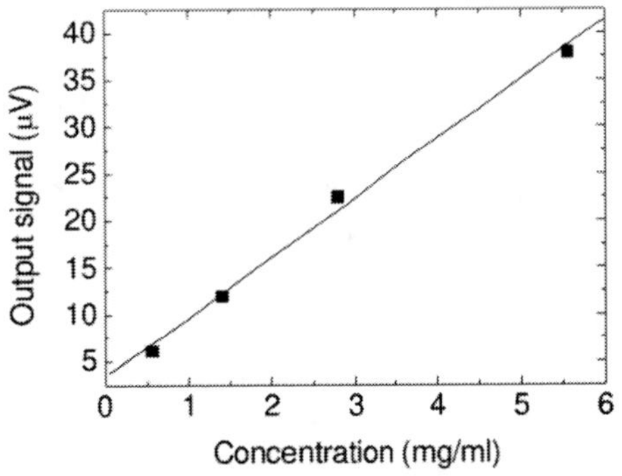

Figure 6. The relationship of the GMR signal to the concentration of ferrofluid.

As a control test, distilled water plugs were pushed into micro channel, and the real time output of bridge voltages were in the range of ±3 μV, which is considered to be system noise. The detection limit of concentration of the ferrofluid is 0.56 mg/ml, and the corresponding signal is 6.2 μV. Because DC magnetic field was applied in this study, a base line drift was observed in measurement. It should be more suitable to apply an AC magnetic field as an induced magnetic field to increase the sensitivity of detection system [4, 11].

4. Conclusions

The GMR sensors based on the [Cu/NiFeCo]×10/ Ta multilayers are applicable to detection of magnetic microbeads and the detectable MyOne beads are about 100 at least. The GMR bridge signal is proportional to the surface coverage of the magnetic beads. The sensitivity of the GMR sensor is inversely proportional to the feature size of the sensor.

The GMR sensor is also capable of detecting the flow of ferrofluid, and the output signals are proportional to the concentration of the ferrofluid. Because of the drift of the zero point in DC magnetic field, an AC magnetic field mode of measurement should be more suitable especially for dynamic detection of ferrofluid.

Acknowledgments

This work was supported by National Natural Science Foundation of China (60971039) and National High Technology Research Program of China (2007AA04Z310).

References

[1] Baselt D R, Lee G U, Natesan M, Metzger S W, Sheehan P E and Colton R J 1998 *Biosensors & Bioelectronics* **13** 731

[2] Wang S X and Li G X 2008 *IEEE Trans. Magn.* **44** 1687

[3] Miller M M, Sheehan P E, Edelstein R L, Tamanaha C R, Zhong L, Bounnak S, Whitman L J and Colton R J 2001 *J. Mag. and Mag. Mater.* **225** 138

[4] Rife J C, Miller M M, Sheehan P E, Tamanaha C R, Tondra M and Whitman L J 2003 *Sensors and Actuators A* **107** 209

[5] Schotter J, Kampb P B, Becker A, Pühler A, Reiss G and Brückl H 2004 *Biosensors and Bioelectronics* **19** 1149

[6] Li G X, Joshi V, White R L, Wang S X, Kemp J T, Webb C, Davis R W and Sun S H 2003 *J. Appl. Phys.* **93** 7557

[7] Graham D L, Ferreira H A, Feliciano N, Freitas P P, Clarke L A and Amaral M D 2005 *Sensors and Actuators B* **107** 936

[8] Ferreira H A, Graham D L, Freitas P P and Cabral J M S 2003 *J. Appl. Phys.* **93** 7281

[9] Shen W F, Liu X Y, Mazumdar D and Xiao G 2005 *Appl. Phys. Lett.* **86** 253901.

[10] Ejsing L, Hansen M F, Menon A K, Ferreira H A, Graham D L and Freitas P P 2004 *Appl. Phys. Lett.* **84** 4729

[11] Pekas N, Porter M D, Tondra M, Popple A and Jander A 2004 *Appl. Phys. Lett.* **85** 4783

[12] Jiang Z, Llandro J, Mitrelias T and Bland J A C 2006 *J. Appl. Phys.* **99** 08S105

Journal of Physics: Conference Series **263** (2011) 012003 doi:10.1088/1742-6596/263/1/012003

Application of magnetic sensors in automation control

Chunhong Hou [1] **and Zhenghong Qian** [2*]

[1] AMETEK Inc., Paoli, PA 19301 U.S.A
[2] Center For Integrated Spintronic Devices (CISD), Hangzhou Dianzi University, Hangzhou, ZJ 310018, China.

E-mail: zqian@hdu.edu.cn

Abstract: Controls in automation need speed and position feedback. The feedback device is often referred to as encoder. Feedback technology includes mechanical, optical, and magnetic, etc. All advance with new inventions and discoveries. Magnetic sensing as a feedback technology offers certain advantages over other technologies like optical one. With new discoveries like GMR (Giant Magneto-Resistance), TMR (Tunneling Magneto-Resistance) becoming feasible for commercialization, more and more applications will be using advanced magnetic sensors in automation. This paper offers a general review on encoder and applications of magnetic sensors in automation control.

1.Overview—control and encoding

In order to achieve motion control, one needs information about the speed and/or the position of the moving or sliding part, which will be referred to as "slider" in linear motor or "rotor" in rotational motion. Such sensing and feedback device, which can be called "encorder" or "tachometer" for speed measurement only, could be purely mechanical. Mechanical signal is generated when a mechanical part touches another one like a mechanical revolution counter or the alarm in a mechanical clock. A device with mechanic contact can generate electric signal when a mechanical part comes in touch with another mechanical part. Shown in Fig. 1 is an example. There is a disk with conductive patterns or pads that are separated by insulating material and a conductive probe or pin. Whenever the disk or the pin rotates to where the pin comes in contact with the conductive pads, a current signal or a pulse can be generated. The signal is feed into a control system, which makes use of the information from the signal to complete certain motion control, either speed control or position control or both. The mechanical feedback device is simple, typically of low resolution but robust and is still being used today in some industry including applications in automobiles. Nevertheless, mechanical feedback devices can't meet the requirements for many other applications. The technology moves on to optical, electronic and magnetic ones.

In general, an encoder can generate signal to measure position and/or speed, which be classified as relative or absolute encoder. A device that generate signal to measure speed only is usually called tachometer. A relative encoder is also referred to as incremental or quadrature encoder, which generates signal about the relative position of a rotor or slider to a reference point. On the other hand, an absolute encoder, which is sometimes also called as parallel absolute encoder, generate signal about the absolute position of a rotor or slider.

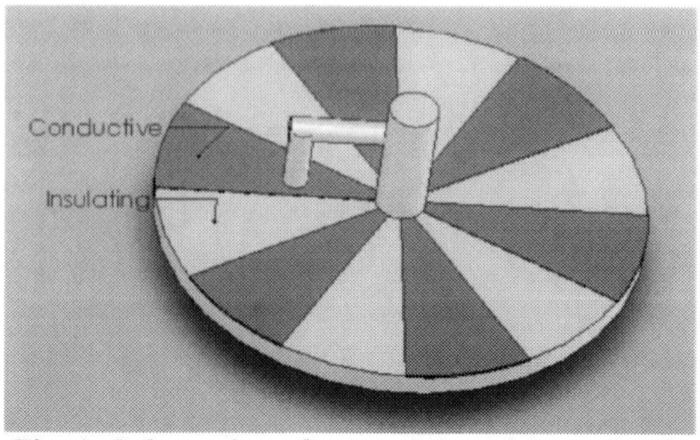

Fig. 1 Schematics of an encoder with mechanic contact.

What shown in Fig. 1 is actually a tachometer. By measuring the time the pin moves between each pad, one can obtain the speed of the rotor, but not the position of the rotor.

Typically two signal tracks and two sensors are used in a relative encoder as shown in Fig. 2 as a simplest example. The two tracks have signal patterns separated at 90 degree apart (quadrature). If we denote the blank area as "0" and the dark area as "1", the coding looks like in Table I, while the pulse sequence looks like in Fig. 3 when the disk rotates in the clockwise direction. The coding will look like in Table II if the disk rotates in counter clockwise direction. From the pulse sequence, one can tell whether the rotation is clockwise or counter-clockwise as well as the speed, if time is recorded, but not the position yet. In order to know the position, another signal is needed as a reference for at least once every rotation. However, if a machine stops, the next time when the machine starts, the feedback system can't tell the machine where the position is until one rotation is completed through the reference point. Note that in this specific example as in Fig. 2, there are only two patterns, blank and dark, in one track. The two sensors can actually generate information about the absolute location of the disk within 90 degree resolution. This coincides with an absolute encoder in this simplest example. If there are more than 2 patterns like blank, dark, blank, dark, etc. on one track, the two sensors can't generate information about the absolute location of the disk.

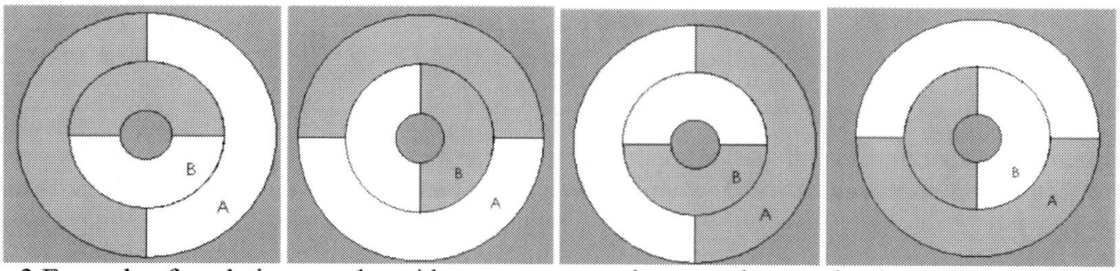

Fig. 2 Example of a relative encoder with two sensors and two tracks, rotating in clock wise direction.

Journal of Physics: Conference Series **263** (2011) 012003 doi:10.1088/1742-6596/263/1/012003

Table I Coding of a relative encoder in Fig. 2 when rotating in clockwise direction.

A	B
0	0
0	1
1	1
1	0

Table II Coding of a relative encoder in Fig. 2 when rotating in counter clockwise direction.

A	B
0	0
1	0
1	1
0	1

In an absolute encoder, usually more than 2 sensors are used and the number of sensors depends on the resolution. The math is simple, resolution=2^N, where N is the number of sensors needed. For a simple case when 3 sensors are used, the coding looks like in Table III. Note that there are different ways of coding. The one in Table III is the so-called Gray coding, where two successive values differ in only one bit, which is more error proof than other ways of coding. In this example, there are 3 sensors, so the resolution is 8. This means one can get as accurate as 45 degree since 360/8=45. Each of the 45 degree area has a unique and therefore absolute code. Compared with the relative encoder, the feedback system made of absolute encoder can always tell a machine where its position is as long as the power is on.

With the introduction completed, we will turn to specific encoder types, optical and magnetic.

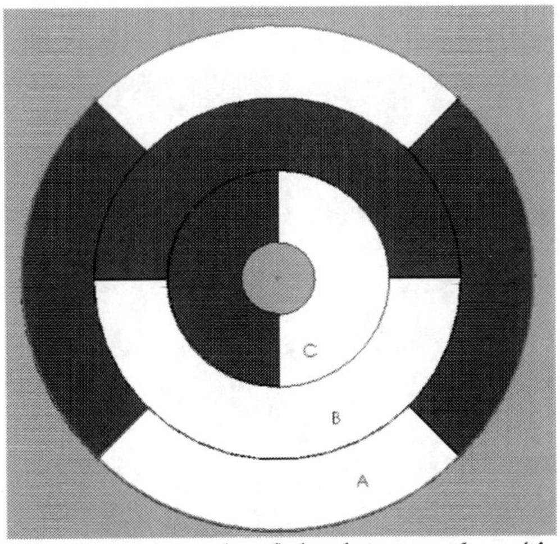

Fig. 3 An example of absolute encoder with three tracks and three sensors.

Table III Gray coding for an absolute encoder with 3 tracks and 3 sensors

	Gray Coding		
Sector Angle	Track 1	Track 2	Track 3
0° to 45°	0	1	1
45° to 90°	0	1	0
90° to 135°	1	1	0
135° to 180°	1	1	1
180° to 225°	1	0	1
225° to 270°	1	0	0
270° to 315°	0	0	0
315° to 360°	0	0	1

2.Optical encoder

An optical encoder is made of a light source, a disk and a receiver. In a transmitting optical encoder, light pass through open slots on a disk as shown in Fig. 4. In a reflective optical encoder, light gets reflected from a disk back to a receiver on the same side of the light source. In both cases, light intensity changes according to the patterns on the disk.

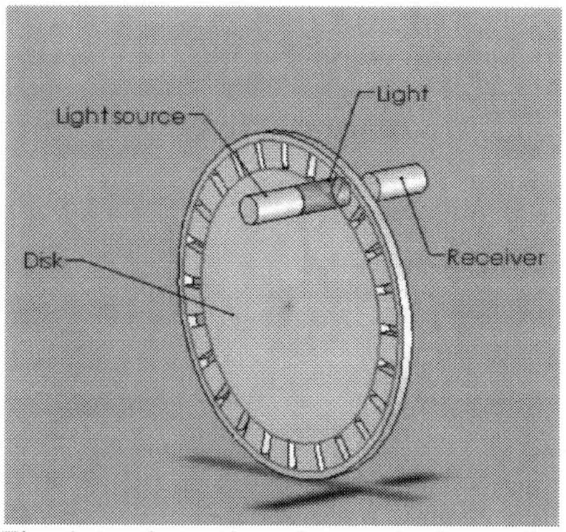

Fig. 4 A schematics of an optical encoder

The typical advantages of optical encoder are high resolution, and inexpensive, though magnetic encoder is catching up in terms of resolution in recent years. The typical disadvantages are prune to shock, harsh environment where dirt and oil could block the light, high temperature, etc.

3.Magnetic sensor and encoder

3.1Hall

Hall sensors have been widely used in motor and control for commutating and encoding.

In brush DC motors, currents are directly commutated by brushes. In brushless DC motors, currents are supplied by a drive. There two types of drives, sinusoidal and trapezoidal. In a typical trapezoidal brushless DC motor, there are three phases that are typically connected in "Y" configuration. Each phase can generates back EMF (Electromotive force) when rotor rotates. The supplied current needs to be on phase with the back EMF in order to generate maximum torque. Typically three magnetic sensors are used to detect the position of the rotor. The information is feed into a drive to trigger current supply into the phases. Most popular sensors for this application are digital Hall sensors. This is shown in Fig. 5.

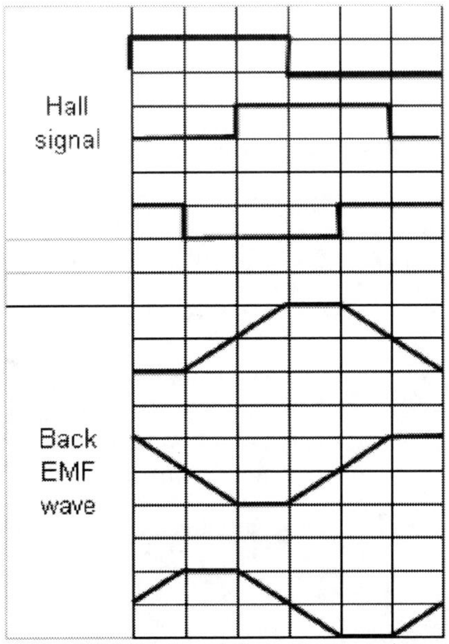

Fig. 5 Hall signal for commutating and back EMF wave of a typical trapezoid motor.

Most motors have large enough size to accommodate the three Hall sensors. For miniature motors, there are cases that the Hall sensors are too large. People have no choice, but go sensor-less, where the drive can pick up the back EMF signal and trigger current supply accordingly. The sensor-less design has its disadvantages. People use this design mostly because there is no other choice. Other types of magnetic sensors like AMR(Anisotropic Magneto-Resistance), GMR or TMR can be intrinsically made small or even integrated into a circuit, whether they will be used for commutating is an uncharted territory. Factors like the overall package size, package compatibility, and cost need to be considered for this application.

Besides being used in commutation, Hall sensors have been widely used in motor encoding for control. Resolution as high as 14 bits or 16384 per revolution has been claimed[1]. An example is shown in Fig. 6, where 8 halls are evenly spaced around a circumference of a circle with the two pole magnet radially magnetized. When the magnet rotates, the 8 halls pick up signal simultaneously. The position of the magnet is obtained by taking the signal from all the 8 halls into consideration for calculation. Another configuration is to have the Halls sitting underneath the two pole magnet as shown in Fig. 7 with the magnet axially magnetized. These types of encoder are not truly digital one. It extrapolates the information from the Halls to "predict" the position in a digital format. If the magnet is magnetized into multiple poles, with the halls at certain location, it is possible to generate truly digital output. Whether it can achieve high resolution depends on the sensitivity of the sensors and the gap between the sensors and the magnet. Even though they are not truly digital encoders, they have been widely used in automation, but the accuracy of this type encoder is sensitive, subject to change in temperature, mounting accuracy, etc. Because of this and other factor like cost, these types of encoder have not found wide application in harsh environment like automobiles and others.

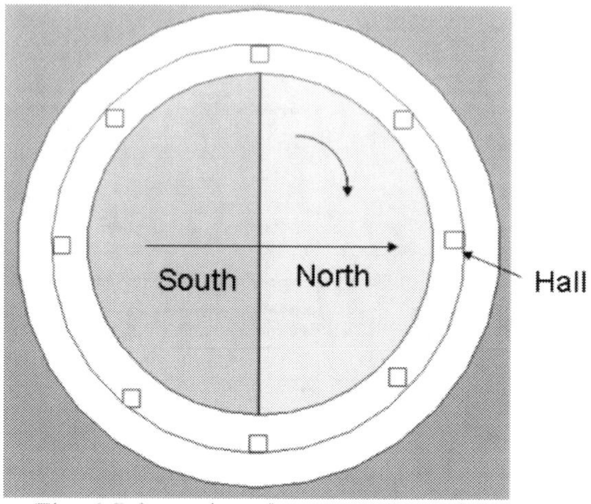

Fig. 6 Schematics of a magnetic encoder with magnet radially magnetized.

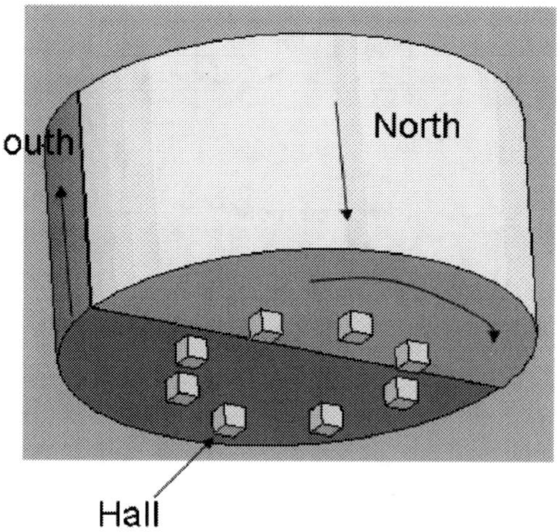

Fig. 7 Schematics of a magnetic encoder with magnet axially magnetized.

3.2 AMR

AMR phenomenon[2], the resistance of a magnetic material varying with regard to direction of the magnetization, follows a formula as:

$$R=R_0+\Delta R \cos^2\theta$$

Where R_0 is the resistance when its magnetization is perpendicular to that of the sensing current; ΔR is the maximum resistance change; θ is the angle between its magnetization and the orientation of the sensing current. The typical AMR effect is about 2-5%, meaning the resistance of a magnetic material could have a difference of 2-5% of its resistance when its magnetization change with respect to that of the sensing current or $\Delta R/R_0 \sim 2\text{-}5\%$. Even though, it is still more sensitive than regular Hall sensors and can be miniaturized.

AMR had not found wide applications until 90's of last century when hard disk drives made a transition from inductive reading to AMR reading technology. Besides that, AMR has been used in MRAM (Magnetic Radom Access Memory) for niche market applications. There is no report of wide application in motor industry due to factors like necessity, cost, etc. Currently commercial AMR sensors are available from Honeywell, NXP semiconductor, Sensitec. AMR sensors in principle can be used to replace the Halls in all previous examples, but one may need to modify circuits to accommodate difference in signal pattern.

3.3 GMR and TMR

GMR[3-4] was discovered in 1988 and won the physics Nobel prize in 2007. It has been widely used in disk drive since late 90's and now starts to be replaced by TMR[5]. The GMR effect comes from spin scattering, where highest resistance is observed when electrons with spin polarized by one magnetic layer move to another magnetic layer of opposite magnetization direction and lowest resistance occurs of the same magnetization direction. The simplest structure is shown in Fig. 7 where two magnetic layers are separated by a metal layer, typically of copper. The resistance formula of this structure is described as

$$R=R_0+\Delta R\,(1-\cos\theta)/2$$

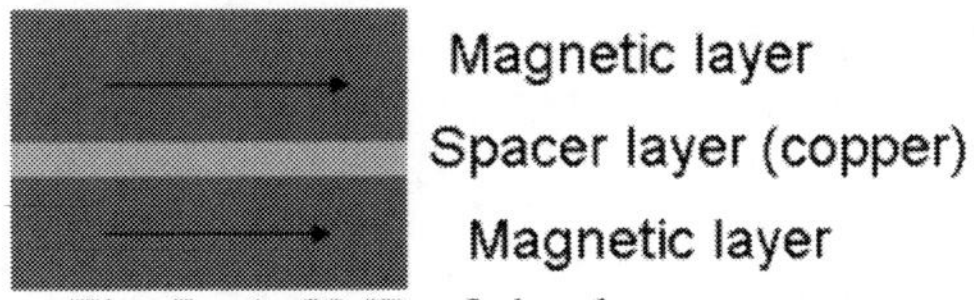

Fig. 7 A GMR of simplest structure.

where R_0 is the resistance when the two magnetic layers have its magnetization in parallel; ΔR is the maximum resistance change; θ the angle between the magnetization of the two magnetic layers. Typical GMR effect is tens of percent, meaning the resistance of a GMR structure could have a resistance change or $\Delta R/R$ up to tens of percent when the magnetization of the adjacent magnetic layers change direction in relative to each other.

TMR was discovered earlier than GMR, but not been widely used until early this century in disk drive. In a typical TMR structure, the (copper) spacer layer is replaced by an insulating layer like Al_2O_3, MgO, etc. In TMR, the sensing current pass through all the layers, often called current perpendicular to plane configuration; while in GMR, the sensing current flow in the plane of the layers, often called current in plane configuration. TMR effect could be as large as hundreds of percent.

The most practical structure of GMR or TMR is the co-called spin valve type in Fig. 8, where one of the magnetic layers, the co-called pinned (magnetic) layer, has its magnetization fixed or pinned by an antiferromagmetic layer or synthetic antiferromagentic layers, at a certain direction and only the magnetization of the other layer, the so-called free (magnetic) layer, reacts to field and change its direction.

Fig. 8 Spin Valve GMR/TMR structure.

Both GMR and TMR have significantly higher sensitivity than Hall sensors. Commercial GMR sensors are available from NVE and Sensitec, but no significant application in motor industry has been reported. In hard disk drives, the tracks are magnetized in either north or south polarity, a true digital format. Currently the linear density in hard drives is over 2000k bpi (bit per inch). This means a bit occupies a length of 12.7 nm or less. If the same linear density can be made feasible in rotary encoder of 10 mm diameter, the resolution would be almost 2.5 millions per revolution and it is truly digital! No encoder can get near to this resolution today, but there are serious challenges to make use of the drive technology in encoders. Still, patents of GMR to be used as an encoding technology have been granted[2-6]. Shown in Fig. 9 is one example. In principle, GMR and TMR can be easily miniaturized and integrated into circuits, and should find applications in miniature motor applications if not for large size motor. The possible issues include cost with GMR or TMR sensor being multiple or ten times or more costly than halls; package compatibility; for example, NVE's GMR sensors have completely different packages from that of the Hall sensors, etc.

One common issue with regard to new magnetic technology like AMR/GMR/TMR is that experts in automation and control may not have much awareness of the new technologies and their superior advantages. For example, these new technologies can offer significantly higher sensitivity, which means the magnets used to produce field to actuate the sensors can be made smaller. This in many applications means low inertia for the rotor or slider, which in turn means higher acceleration, more precision control, etc.

4.Conclusions

In today's automation and control industry, optical and Hall sensor based magnetic technologies dominate the market, while mechanic still plays some role. But the new magnetic technologies like AMR/GMR/TMR have yet to be widely used in spite of super advantages in sensitivity, temperature stability, etc. due to factors like lack of awareness by experts in automation and control, cost, connection and package compatibility, etc. However, there are some applications, like in miniature motors, high end products that require high accuracy and low inertia, etc., where regular Hall sensors can not be used due to factors like limited space or signal sensitivity, temperature stability, etc. This is where AMR/GMR/TMR can find applications.

References:

[1] http://www.austriamicrosystems.com/
[2] W. Thomson, Proc. R. Soc. London 8, **546** (1857)
[3] Baibich M N, Broto J M, Fert A, Nguyen Van Dau F, Petroff F, Binasch G, Grünberg P, Saurenbach F, and Zinn W 1988 *Phys. Rev. Lett.* **61** 2472–2475
[4] G. Binasch, P. Grünberg, F. Saurenbach, and W. Zinn 1989 *Phys. Rev. B* **39**, 4828–4830
[5] M. Julliere 1975 *Phys. Lett.* **54A** 4 225-226

[6] Andreas D, Marcus B, Hubert G, Karl-heinz L, Nikolaus L, Rolf S, and Guenther M 2009 High resolution magnetic encoder Patent No. 7,612,555
[7] Fries D 2005 Giant magnetoresistance based nanopositioner encoder Patent No. 6,940,277,
[8] Lohberg P 2004 Linear distance sensor and the use thereof as actuator for motor vehicles Patent No.6,823,725,
[9] Heinze F, Schäfer D 2008 Position transmitter and method for determining a position of a rotating shaft Patent No.7,466,125,
[10] Yao M 2008 Vehicle wheel speed and acceleration sensor and method for calculating the same Patent No.7,405,554,

Novel spintronic device: terahertz magnon-photon laser

B G Tankhilevich and Y Korenblit

Terahertz Technologies, LLC; 536 North Civic Drive, #A, Walnut Creek, CA (US)

E-mail: borist@pacbell.net

Abstract. A novel method of generating THz radiation is proposed. The method is based on pumping of non-equilibrium electrons into the upper (spin-down) subband of spin-polarized half-metallic ferromagnets or ferromagnetic semiconductors. The electrons rapidly emit magnons with THz frequency, pass into highly excited states of the spin-up subband, and fall into the ground state due to interaction with the equilibrium spin-up electrons or emitting optical phonons. The mechanism of magnon generation is similar to a three-level laser, and at a critical pumping intensive magnon lasing begins. Merging of two magnons generates a THz photon. Thus, a magnon laser becomes a THz photon laser, the generated power being of orders of milliwatt vs. several microwatt power generated by contemporary devices. The generated frequency is material dependent, and can be changed by tuning the magnetic field and the bias. The proposed device may have many applications in different fields including medicine (cancer surgery), molecular imaging, security problems etc.

1. Introduction

The THz region of the electromagnetic spectrum defined by the frequency range of 0.3 to 10 THz have recently become the subject of heightened activity in the physic and engineering communities because of promising new applications in semiconductor physics, medical, space and defence industries (for a review see [1]). As argued in the above review, the lack of a high-power, low-cost THz source is the most significant limitation of modern THz systems. In recent years some new methods were proposed for THz wave generation [2],[3],[4],[5]. However, a powerful compact tunable narrow-band source of electromagnetic THz radiation is still lacking.

In this paper a novel method of generating THz radiation, a THz magnon-photon laser is proposed, which is capable of generating narrow-band THz power of orders of milliwatt [6]. The proposed source utilizes spin-polarized ferromagnetic materials like half-metals and ferromagnetic semiconductors. In both of these materials the conduction band is split by the exchange interaction into two subbands with opposite spin orientation, and only electrons states in the lower subband ("spin up" majority electron states) are occupied at zero temperature. The exchange gap Δ is of order of hundreds meV.

Nonequilibrium electrons pumped into the upper subband ("spin down" minority electron states) rapidly emit magnons, with a large wave vector $q \approx \hbar^{-1}(2m\Delta)^{1/2}$, where m is the electron effective mass. The frequency of these magnons fall into the THz region. For instance the classical ferromagnetic semiconductor EuO (Curie temperature, Tc = 69°K) the gap is Δ= 0.6 eV. With $m \approx m_0$, where m_0 is the free electron mass, and the magnon stiffness, D =10.8 meV·A^2, the energy of the excited magnons *is* $\hbar\omega = Dq^2$ =2.1 meV, and the frequency $f_m = (2\pi)^{-1}\,\omega$=0.54 THz. In half-metals ($T_c \geq$ 300 K) Δ is the distance between the Fermi energy and the bottom of the minority carrier band, and it lies in the region 0.1–0.4 eV. The stiffness D is of order of 100-200 meV· $Å^2$. Thus, the frequency of the excited magnons should be several THz.

Merging of two magnons with frequency f and wave vectors **q** and -**q** generates a photon with frequency 2f. This process is a process reversal to the well known process of parametric magnon generation by electromagnetic radiation. At critical pumping currents $j_c \approx 10^4 - 10^5$ $Å/cm^2$ the magnon lasing process begins, and the device should generate high-power narrow-frequency THz radiation. The device is of nano scale and is capable of large scale productions. The device can also work at room temperature if half metals are employed.

2. Dynamics of a magnon laser

The system of equations which govern the behavior of the pumped electron, $f_\downarrow(\mathbf{p})$, and magnon, $N(\mathbf{q})$ distribution functions can be written as

$$\partial N(\mathbf{q})/\partial t = [1+N(\mathbf{q})]\,\Gamma_e(\mathbf{q}) - [N(\mathbf{q}) - N^0(\mathbf{q})]\,\Gamma_{ms}(\mathbf{q}) \tag{1}$$

$$\partial f_\downarrow(\mathbf{p})/\partial t = g(\varepsilon_\mathbf{p}) - f_\downarrow(\mathbf{p})\,\gamma_{em}(\mathbf{p}) \tag{2}$$

Here $g(\varepsilon_\mathbf{p})$ is the generation function of the spin-down electrons. $\Gamma_e(\mathbf{q})$ is the relaxation rate of magnons in collisions with electrons

$$\Gamma_e(\mathbf{q}) = 4\pi\,\hbar^{-1} I^2 S\, v_0 \int d^3p (2\pi\hbar)^{-3}\, \delta(\varepsilon_{\mathbf{p}\downarrow} - \hbar\omega_\mathbf{q} - \varepsilon_{\mathbf{p}-\mathbf{q}\uparrow})\, f_\downarrow(\mathbf{p}), \tag{3}$$

where v_0 is the unit cell volume, and I is the s-d (s-f) exchange interaction.

Γ_{ms} is the relaxation rate of the magnons to the equilibrium value $N^{(0)}$, which includes the magnon-magnon scattering, Γ_m, scattering of magnons on the surface, and the scattering of magnons by the majority spin-up electrons. Note that in both half-metals and ferromagnetic semiconductors the last scattering is relatively small, since one-magnon processes are forbidden by the energy conservation law.

γ_{em} is the electron-magnon relaxation rate

$$\gamma_{em}(\mathbf{p}) = 4\pi\,\hbar^{-1} I^2 S\, v_0 \int d^3q (2\pi\hbar)^{-3}\, \delta(\varepsilon_{\mathbf{p}\downarrow} - \hbar\omega_\mathbf{q} - \varepsilon_{\mathbf{p}-\mathbf{q}\uparrow})\, (1+N(\mathbf{q})). \tag{4}$$

The magnon emission is described by the first term in (1). It follows from the energy conservation law that if $\varepsilon_\mathbf{p} = p^2/2m$ is smaller thanΔ, the wave vectors q lie in a smooth interval $q_1 \leq q \leq q_2$, where $q_{1,2} = \hbar^{-1}(p_0 \pm p)$, $p_0 = (2m\Delta)^{1/2} \gg p$.

After the emission of magnons the electrons pass into highly excited states of the spin-up subband. Since the ionicity in ferromagnetic semiconductors like EuO is large, the electrons in this state will most probably emit optical phonons and fall into low-energy states of this subband, see figure 1.

In half-metals this should happen due to the strong interaction of the high-energy electrons with the spin-up electrons at the Fermi surface. As a result the absorption of magnons by the spin-up electrons in the highly excited states is strongly suppressed, and the system acts as a three-level laser.

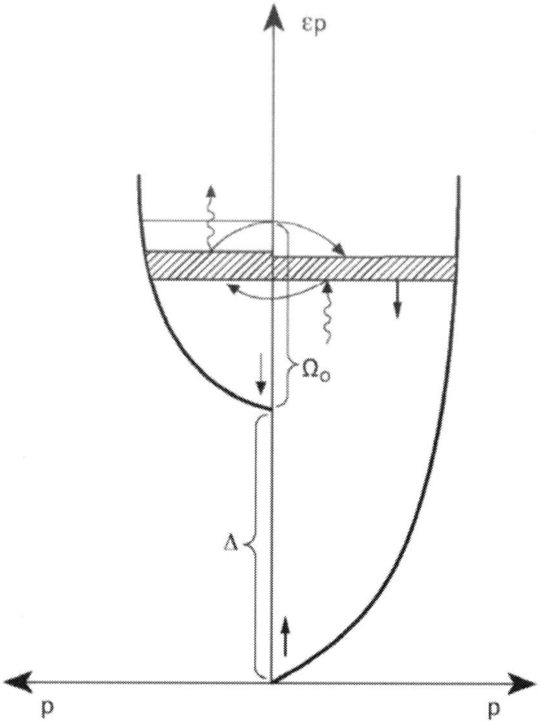

Figure 1. This figure illustrates the process of strong electron–magnon interaction (comparatively with electron-electron or with electron-phonon interaction) wherein a nonequilibrium electron put in the upper subband with spin down rapidly emits a magnon with a large wave vector.

The electron-magnon relaxation rate γ_{em} is very large, of order $3 \cdot 10^{12} - 10^{13}$ sec^{-1}. Thus, after a very short time of order γ^{-1}_{em} the electron distribution function according to (2), reaches the value

$$f_{\downarrow}(\mathbf{p}) = g(\varepsilon_{\mathbf{p}})\,(\gamma_{em}(\mathbf{p}))^{-1} \tag{5}$$

It follows from this equation and (3) and (4):

$$\Gamma_e(\mathbf{q}) = \int d^3p\, g(\varepsilon_{\mathbf{p}})\, \delta(\varepsilon_{\mathbf{p}\downarrow} - \hbar\omega_{\mathbf{q}} - \varepsilon_{\mathbf{p-q}\uparrow}) Z^{-1}(\mathbf{p}), \tag{6}$$

with

$$Z(\mathbf{p}) = \int d^3q\, \delta(\varepsilon_{\mathbf{p}\downarrow} - \hbar\omega_{\mathbf{q}} - \varepsilon_{\mathbf{p-q}\uparrow})\,(1 + N(\mathbf{q})). \tag{7}$$

These equations together with (1) give a complicated integro-differential equation for $N(\mathbf{q},t)$. We solve it in the limits of small and large t assuming that the temperature, T, is low, $kT \ll \hbar\omega(\mathbf{q}_0)$, and therefore $N(\mathbf{q}, 0) = \overset{0}{N}(\mathbf{q})$ is much smaller than unity.

Consider not too large times, when $N(\mathbf{q})$ is also small, $N(\mathbf{q}) \ll 1$. Then (1) transforms into a simple differential equation

$$\partial N(\mathbf{q}) / \partial t = \Gamma_e^*(\mathbf{q}) - [N(\mathbf{q}) - N^{(0)}(\mathbf{q})]\Gamma_{ms}(\mathbf{q}) \tag{8}$$

with Γ_e^* given by (6), in which the functional Z does not depend on $N(\mathbf{q})$. The solution of (8), with the initial condition $N(\mathbf{q}, 0) = N^{(0)}$, is

$$N(\mathbf{q}, t) = (N^{(0)}(\mathbf{q}) - \Gamma_e^*(\mathbf{q})(\Gamma_{ms}(\mathbf{q}))^{-1}) \exp(-\Gamma_{ms}(\mathbf{q})t) + \Gamma_e^*(\mathbf{q})(\Gamma_{ms}(\mathbf{q}))^{-1} \quad (9)$$

This solution holds only at such t, when the condition $N< 1$ is satisfied. It is seen that if the ratio $\Gamma_e^*(\Gamma_{ms})^{-1}$is less than unity, N changes from $N = N^{(0)}$ at $t = 0$ till

$$N(\mathbf{q})= \Gamma_e^*(\mathbf{q})(\Gamma_{ms}(\mathbf{q}))^{-1} \quad (10)$$

at t larger than$(\Gamma_{ms})^{-1}$, being always small. Thus, in this case only spontaneous emission of magnons is important. In the stationary state given by (10), $N(\mathbf{q})$ increases linearly with the pumping.

If $\Gamma_e^*(\Gamma_{ms})^{-1}$exceeds one, the solution of (9) is valid only at $t \ll (\Gamma_{ms})^{-1}$, when N remains small. When N reaches a value of order unity, the stimulated emission dominates. Thus, the threshold pumping for a fast increase of $N(\mathbf{q})$ caused by the stimulated emission can be written as

$$\Gamma_e^* = \Gamma_{ms} \quad (11)$$

The same condition follows from the arguments borrowed from the theory of optical lasers. The larger the pumping the less is the time delay required for the stimulated emission to become dominant.

We calculate $\Gamma_e^*$for two types of pumping:

$$g(\varepsilon_p) = g_0\varepsilon\delta(\varepsilon-\varepsilon_p).$$
$$g(\varepsilon_p \leq \ \varepsilon) = g_{0;} \qquad g(\varepsilon_p > \varepsilon) = 0. \quad (12)$$

One obtains from (6) and (7) that Γ_e^* is the same for both types of pumping given by (12)

$$\Gamma^*e = g_0\,\varepsilon\,(4\Delta)^{-1}. \quad (13)$$

It follows then from (11) that the threshold pumping is given by

$$g_0 \equiv g_{th} = 4\Delta\,(\varepsilon)^{-1}\,\Gamma_{ms} \quad (14)$$

If the pumping exceeds the above threshold, the behavior of $N(\mathbf{q})$ depends on the anisotropy of the generation.

Suppose first that the generation is isotropic, and N(q) does not depend on the direction of **q**. Then the function Z increases with the increase of the pumping in such a way, that Γe ia always smaller than Γ_{ms}. In other words, it is the feedback relationship between the magnon emission rate and the population of the subband with spin down (see (2)), which leads to the steady state at sufficiently large t.

When the pumping exceeds g_{th} the stimulated emission of magnons becomes important, and the number of magnons increases nonlinearly with pumping. At pumping $g_{0.} > g_c = \Delta^{/2}\varepsilon^{-1/2-}g_{th}$ the system is in a steady state, with N(q) increasing exponentially with pumping in a smooth range of magnon wavelengths q near q_1, wherein the range of magnon wavelength q decreases exponentially with the increase of pumping.

The t-dependence of N is quite different, if the magnon emission is anisotropic, i.e. $N(\mathbf{q},t)$ increases mainly at **q** in the vicinity of some $\mathbf{q}^*$ (there may exist several such optimal vectors). Then the function Z, which is an average of $N(\mathbf{q},t)$, is not very sensitive to the value of $N(\mathbf{q}^*,t)$, and the feedback relation discussed above is not essential. Therefore, Γ_e can be considered as time independent even in the case, when $N(\mathbf{q}^*,t)$ is large. The solution of (1) is:

$$N(\mathbf{q}^*, t)= N^{(0)}(\mathbf{q}) + (\Gamma_e(1+ N^{(0)})\,(\Gamma_e-\Gamma_{ms})^{-1})\,[\exp((\Gamma_e-\Gamma_{ms})t) -1]. \quad (15)$$

Thus, $N(\mathbf{q}^*, t)$ increases exponentially with time, if g_0 exceeds g_{th}. At some time, when N becomes very large the feedback may cause the system to enter into a steady state also at anisotropy conditions.

One can get strong anisotropy of the magnon emission using two magnon mirrors [6] with an active region in between. The magnon mirrors serve as a magnon resonant cavity, and when g_0 exceeds g_{th}, the magnons, which move perpendicular to the magnon mirrors, are mainly emitted, thus leading to an exponential increase of their number with time.

3. Generation of THz radiation

The interaction of magnons with electromagnetic radiation was considered in [7]. Merging of two magnons with wave vectors **q** and **q**' generates a photon with wave vector

$$\mathbf{k} = \mathbf{q} + \mathbf{q}', \tag{16}$$

and with frequency ν_k equal to

$$\omega_{\mathbf{q}} + \omega_{\mathbf{q}'} = \nu_{\mathbf{k}} = c\mathbf{k} \tag{17}$$

where c is the light velocity. It follows from these conservation laws that k is much smaller than q, i.e. $\mathbf{q} = -\mathbf{q}'$.

The probability of transformation of two magnons in a photon is the greatest, when **q** is along the magnetization. We consider, therefore a magnon laser, in which just these magnons are mainly generated. Then the evolution of the number of generated photons, $n(t)$, with time is given by the equation:

$$dn/dt = WN^2(t)\,[n(t) + 1] - Wn(t)\,[N(t) + 1]^2 - n(\tau_{ph})^{-1}. \tag{18}$$

Here W is the the photon generation probability, which is proportional to the square of the magnetization and is of order $10^7\ s^{-1}$, τ_{ph} is the photon relaxation time. The main relaxation mechanism is the absorption of photons by the conduction electrons.

Suppose that $N(t)$ increases exponentially with time

$$N(t) \propto \exp(\beta t). \tag{19}$$

Then (18) can be solved analytically in two limiting approximations.

3.1. $WNt_{ph} \ll 1$.

Equation (18) can be rewritten as

$$dn/dt + n(t_{ph})^{-1} = WN^2(t). \tag{20}$$

The solution of this equation is:

$$n(t) = W\,t_{ph}\,N^2(t)\,(2\beta\,t_{ph} + 1)^{-1}. \tag{21}$$

In this regime, which is realized, when N is not too large, but, of course, it is much larger than unity, n may be smaller than 1, if WN^2t_{ph} is small, or larger than 1, if the opposite inequality holds.

3.2. $WNt_{ph} \gg 1$.

Then equation (18) transforms into

$$dn/dt + 2WN(t)\,n(t) = W\,N^2(t) \tag{22}$$

One gets

$$n(t) = N(t)/2 + const. \qquad (23)$$

Thus, when the magnon lasing takes place, the number of generated THz photons increases exponentially with time.

3.3. Tunability of the generated radiation

In an external magnetic field H, the spin-wave spectrum for high-frequency magnons is

$$\hbar\,\omega_q = (\hbar\,\omega_0 + Dq^2) \qquad (24)$$

Here $\hbar\,\omega_0 = g_L\mu_B H$, where g_L is the Lande g-factor ($g_L \approx 2$) and μ_B is the Bohr magneton. Hence, the frequency of the generated radiation is

$$f = 2\omega_q/2\pi = \pi^{-1}(\omega_0 + \hbar^{-1}Dq^2) \qquad (25)$$

Thus, the magnetic field-induced tunability, t_H, defined as

$$t_H = \partial f_r / \partial H \qquad (26)$$

is

$$t_H = g_L\mu_B/\pi\hbar = 0.056 THz/T. \qquad (27)$$

The frequency of the radiation can be tuned also by the bias. As mentioned above, at high pumping levels, the generated magnons have wave-vectors, which are close to q_1, i.e., the frequency of the excited magnons are

$$\omega_q = 2mD\hbar^{-2}[\Delta - (\Delta\varepsilon)^{1/2}]. \qquad (28)$$

ε depends on the bias. Therefore ω_q and, hence, the frequency of the generated radiation $f = 2\omega_q/2\pi$ can be tuned by changing the bias.

3.4. Theory vs experiment

In a recent paper [8] THz generation by current induced spin injection into a Permalloy film was observed. The current density was of the order of 10^9 A/cm^2. In Permalloy, unlike half-metals discussed above, the density of states of both majority and minority electrons are finite at the Fermi surface, i.e. there is no exchange gap. Therefore, generation of radiation due to direct transitions of the injected electrons across the exchange gap, considered in [5], does not happen. It seems that the only mechanism, which can lead to THz radiation at the above experimental conditions is that, considered in this paper and in [6]. The exchange shift of the spin-up and spin-down sub-bands in Permalloy is 0.2 eV, and therefore the frequency of the excited magnons, and, hence, of the generated radiation should indeed be of order of several THz. However, since there is no exchange gap in Permalloy, there is a strong absorption of magnons by spin-up (majority) electrons, and there is no magnon lasing.

4. Applications and Commercialization

4.1. Medicine (cancer surgery)

Imaging system based on the proposed Terahertz Magnon-Photon Laser would enable a real-time image-guided cancer surgery.

One of the first considerations of the issue of using micro millimeter to centimeter wavelengths for imaging human tissue is the tissue resolution obtainable. This is of course related to the wavelength in tissue (not in air). Thus for 30 mm wavelengths in tissue of dielectric coefficient between 30 and 60 the wavelength is about 5 mm (wavelength/ √ dielectric coefficient). The resolution for terahertz frequencies is 100 times less but the dielectric coefficient is expected to be also much less resulting in an expectation of effective tissue wavelengths of about 200μm. When one considers the question of the depth of interrogation that plagues optical or EM field imaging, it is important to note that the tissue linear attenuation is much less than the measured attenuation in deionized water and for 1 terahertz attenuation is less by a factor of about 5 from attenuation in the near infra-red. For terahertz frequencies we are dealing with attenuations of 100 to 120 per cm [9] which in simple terms is an attenuation of more than 10,000 1 mm thick. But the contrast between muscle and fat should be over 10:1. Thus it is not surprising to note the high contrast in transmission images of meats and foods using 0.8 terahertz frequencies from the recent work of Fitzgerald and co-workers [10]. In sum, the success with NIR reflection and transmission imaging of breast tissues in vivo has been encouraging in view of the scattering and attenuation suffered by the photons. In addition, from optics differences in tissue types will be reflected in differences in the index of refraction (square root of dielectric coefficient), thus the interfaces between tumor tissue and normal tissue in the sub millimeter range are expected to give image contrast differences.

The maximum terahertz output power from the photo mixer used in [11] was about 10 microwatt at the low-frequency end (100 GHz), just above 1 microwatt at 1 THz, and then began to fall off, thus it was not tunable at all. Thus, as compared with the THz imaging system used in [12] the proposed Terahertz Magnon-Photon Laser imaging system should be able to obtain THz imaging of the border between the cancerous and healthy tissue with much higher spatial resolution (due to tunability) and would also minimize the noise (due to high power output).

Therefore, the proposed Terahertz Magnon-Photon Laser imaging system would provide a truly enabling tool capable of aiding a surgeon in immediately identifying residual cancer after the main tumor has been removed, thus minimizing the need for additional surgical procedures.

4.2. Molecular imaging

THz frequency range allows interrogation of unique molecular vibrations. So far applications [13] THz molecular imaging & spectroscopy have been limited by insufficient power and poor tunability of the generated THz radiation. THz Spectrometer based on the proposed Terahertz Magnon–Photon Laser would enable to overcome these shortcomings by generating almost coherent sufficiently powerful tunable THz beams, which would make it possible to see and differentiate between very narrow resonant vibration modes. Thus, the proposed Terahertz Magnon–Photon Laser Spectrometer would enable bio sensing as means of identification different bio-molecules. This is especially important for the huge pharmaceutical market, which is estimated to be upwards of US $680 Billion (2009 estimates).

4.3. Security applications

On Christmas day 2009, Umar Farouk Abdulmutallab, 23, allegedly concealed in his underpants a package containing nearly 3oz of the chemical powder PETN (pentaerythritol tetranitrate). He also carried a syringe containing a liquid accelerant to detonate the explosive.

After that act of terrorism, the US Transportation Security Administration ordered $165m-worth of scanners, using both millimeter and X-ray technology, from L-3 Communications. Each full body-scanner costs around $200,000.

However, the explosive device smuggled in the clothing of the Detroit bomb suspect would not have been detected by body-scanners set to be introduced in British airports according to The Independent [14].

Indeed, since the attack was foiled, body-scanners, using "millimeter-wave" technology and revealing a naked image of a passenger, have been touted as a solution to the problem of detecting explosive devices that are not picked up by traditional metal detectors – such as those containing liquids, chemicals or plastic explosive.

But, tests by scientists in the team at Qinetiq showed the millimeter-wave scanners picked up shrapnel and heavy wax and metal, but plastic, chemicals and liquids were missed. If a material is low density, such as powder, liquid or thin plastic – as well as the passenger's clothing – the millimeter waves pass through and the object is not shown on screen.

The proposed Terahertz Magnon-Photon Laser Scanner imaging system used to generate high power tunable THz radiation would be able to detect the sub millimeter chemical powder PETN (pentaerythritol tetranitrate) used in the recent terrorist act.

5. Conclusion

We have shown that when spin-down electrons are pumped into ferromagnetic semiconductors or half-metals, one can achieve the conditions for magnon lasing. The dipole-type magnon-magnon interaction results in generating of THz photons. Thus, the proposed devise is a magnon-photon laser. The generated frequency is material dependent and choosing the proper material, THz frequencies occupying the whole THz region from 0.1 THz to 10 THz can be obtained. The critical pumping for lazing depends on the magnon decay, mainly on the exchange magnon-magnon scattering. Hence, to lower the critical pumping, one should work at temperatures much smaller than Tc. Another reason to lower the temperature is the need to use ferromagnets with highly polarized electrons.

There exists now a large variety of half-metals with T_C higher than the room temperature. E.g., Co2FeSi is a half-metal with Tc = 1100 K [15]. This offers the possibility to get intensive THz radiation at room temperature. In ferromagnetic semiconductors T_C is smaller than the room temperature, and therefore they can be used as THz generators only at low temperatures. But in semiconductors the non-equilibrium electrons may be pumped not only by current but also by optical excitation. The last method would reduce the absorption of the radiation by the electrons and increase, therefore, the efficiency of the device.

References

[1] Ferguson B and Zhang X 2002 *Nature Mat.* **1**, 26
[2] Koshelev A E and Bulaevskii L N 2008 *Phys. Rev.* B **770** 014530
[3] Lin S and Hu X 2010 *arXiv* **1001** 5119 v1
[4] Ryzhii V et al 2009 *J.Appl. Phys* **106** 084507
[5] Kadigrobov A et al 2004 *Europhys. Lett.* **67** 948
[6] Korenblit Y and Tankhilevich B 2008 Generation of terahertz waves *US Patent* **7,430,074**; 2008 Tunable generation of terahertz radiation *US Patent* **7,440,178**; 2008 Method and apparatus for generating terahertz radiation with magnon gain medium and magnon mirror *US Patent* **7,471,449**; 2009 Magnon laser *US Patent* **7,508,578**; 2010 Modulation of terahertz radiation *US Patent* **7,706,056**
[7] Kaganov M I and Tsukernik V M 1960 *Sov. Phys.-JETP* **37** 587
[8] Chigarev S G et al 2010 *arXiv* **1004** 5248 v1
[9] Siegel P 2004 Terahertz technology in biology and medicine. *IEEE Trans. Microwave Theory Techniques* **52** 2438
[10] Fitzgerald et al 2005 *The 18th Annual Meeting of the IEEE* p 120 – 121
[11] Hindle F, Cuisset A, Bocquet R and Mouret G 2007 Continuous-wave terahertz by photomixing: applications to gas phase pollutant detection and quantification *Comptes Rendus Physique* **10** 1016.
[12] Fitzgerald A J, Wallace V P, Jiménez-Linan M et al 2006 Terahertz pulsed imaging of human breast tissue *Radiology* May 2006 **239** No. 2 533-540
[13] Schmuttenmaer C A 2004 *Chemical Reviews* **104** 1759
[14] Merrick J 2010 January 3 Are planned airport scanners just a scam? *The Independent* (UK)
[15] Wurmel S et al 2005 *Phys. Rev.* B **72** 184434

Fabrication and Analysis of PM-biased Spin-valve Sensors

Ru Bai[1*], Zhenghong Qian[1], Yucheng Sun[1,2]
1. Center for Integrated Spintronic Devices, Hangzhou Dianzi University, Hangzhou 310018, China
2. College of Materials Science and Engineering, Sichuan University, Chengdu 610064, China.

E-mail: bairu@hdu.edu.cn

Abstract: GMR spin-valve sensors are fabricated and analyzed, where a pair of CoPt permanent magnets (PM) are attached to two ends of the sensor elements to suppress the hysteresis. The size dependence of the spin-valve sensors is investigated by measuring GMR spin-valve sensing elements patterned with different lengths and widths. The results show that the GMR ratio reduces and the saturation field increases respectively with decreased length of the spin-valve sensor. And the saturation field also decreases with increasing width due to a stronger demagnetization effect. The experimental results are in good agreement with the theoretical prediction.

Keywords: Unshielded spin valve; Transfer curve; Linearity; GMR ratio

1. Introduction

Since the pioneering work by Dieny et al. in 1991 [1], Spin valve has proved to be one of the best materials applied in a variety of magnetic devices, e.g. read heads [2] due to higher sensitivity and better linearity compared to conventional Hall-effect sensing elements and anisotropic magnetoresistive (AMR) sensing materials [3-5]. For read head applications, obtaining much higher-sensitive sensors is always one of the key objections to meet the requirement of continuously-increased areal storage densities. In order to achieve higher resolution reading, it is common to place the spin-valve sensing elements through insulator gaps between two magnetic shields, which can eliminate the external field influence on the read signal [6]. However, the spin-valve sensing elements also can be used in many other areas, such as magnetic media readers (e.g. credit card readers), and field sensors etc.. Theoretically, the spin-valve materials are intrinsically linear if the hysteresis is suppressed. How to decrease or eliminate the hysteresis of spin-valve sensor is vital to making excellent linear devices. In this work, we will describe the fabrication and analysis of a GMR spin-valve sensor, of which the hysteresis is suppressed by placing a pair of permanent magnets (PM) on both ends of the sensor, as shown in Figure 1.

The pattered spin-valve sensing elements exhibit different characteristics as compared with the sheet film due to size effect. Herein we will briefly describe the main fabrication process for the GMR spin-valve sensor firstly. Then, we will investigate the dimension dependence of the sensor performance by measuring spin-valve sensing elements patterned with different lengths and widths.

2. Experimental

2.1 Fabrication

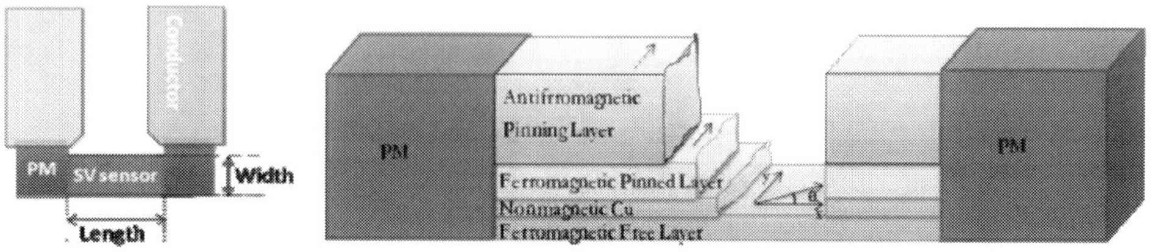

(a) (b)

Figure 1 Schematic configuration of a) GMR spin-valve reader with magnet bias, and b) the cross section of PM-biased, spin-valve sensing element.

The fabrication of the unshielded GMR spin-valve sensor employs the modern manufacture technology, similar to that of spin-valve read head of hard disk drivers, but much simpler. The main process steps, as shown in Figure 2, are explained as below: a) Deposite GMR spin-valve materials on silicon substrate after depositing a SiO_2 insulating layer, and then deposite MRS protect layer; b) Spin-coat photoresist 1; To make the resist lift-off feasible, an undercut resist structure is used for the following PM deposition [7]; c) Pattern spin-valve layer and protect layer by photo and IBE techniques; d) Deposit PM (CoPt used in this case); e) Lift off photoresist 1 and chemical-mechanical polishing (CMP) the surface; f) Spin-coat photoresist 2 and deposite conductor layer; g) Lift off to form touch conductor on the top of the PM.

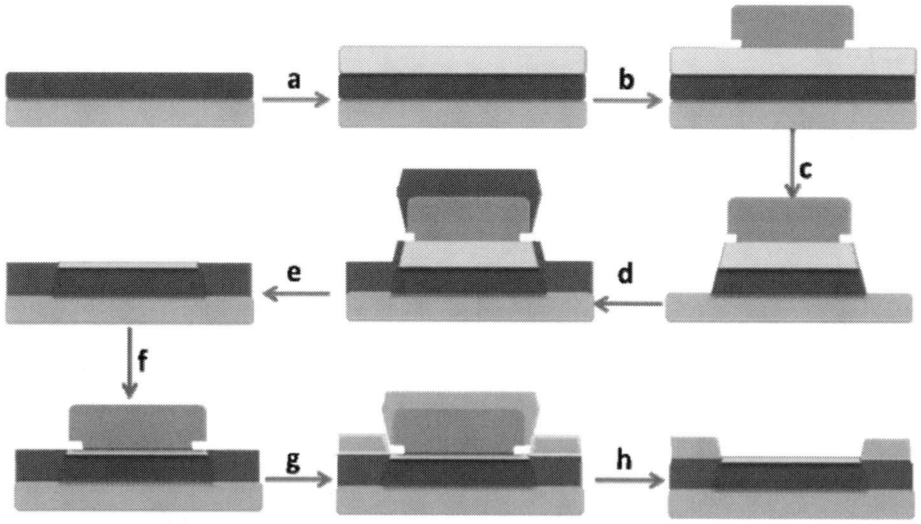

Figure 2 Main process steps for fabrication of the unshielded GMR spin-valve sensor

2.2 Analysis

The experimental GMR transfer curves for the spin-valve sheet films and the patterned spin-valve sensing elements are measured with a magnetic tester, and the experimental curves are compared with the theoretically simulated transfer curves.

3. Results and Discussion

The materials used in the GMR spin-valve sensors in this work (as shown in Figure 1b) has a configuration of NiFeCr(45Å)-NiFe(55Å)-CoFe(15Å)-Cu(21Å)-CoFe(24Å)-Ru(10Å)-CoFe(22Å)-PtMn(250Å)-NiFeCr(70Å). Figure 3 gives the transfer curve of the sheet GMR spin-valve materials, showing excellent performance with a high GMR of about 9.9%.

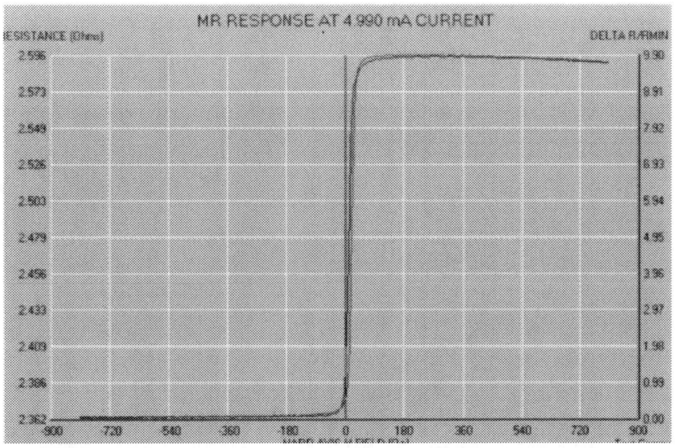

Figure 3 GMR properties of the sheet spin-valve.

Shown in Table 1 are fabricated spin valve sensing resistors with different dimensions.

Table 1 Length and width of tested GMR spin-valve devices

Device 1		Device 2	
Length	Width	Length	Width
14.0	2.5	2.9	1.8

Figure 4 shows respectively the experimental GMR transfer curves and the theoretic simulation for the Device 1. The tested sensor (Device 1) has a dimension of 14 um length and 2.5 um width, and is biased with a pair of 400 Å CoPt permanents (PM). A comparison indicates that the experimental transfer curve (Figure 4a) is roughly in agreement with the normalized simulation result (Figure 4b). It should be noted that, with biased by CoPt permanent magnets, the GMR spin-valve sensors theoretically should have no hysteresis, since the biased field generated by the patterned CoPt permanent magnets could induce a single domain structure in free layer and thus suppress the hysteresis of GMR spin-valve sensing elements. However, as can be seen from Figure 4a, the experimental transfer curve still has little hysteresis, indicating the unperfect bias in Device 1.

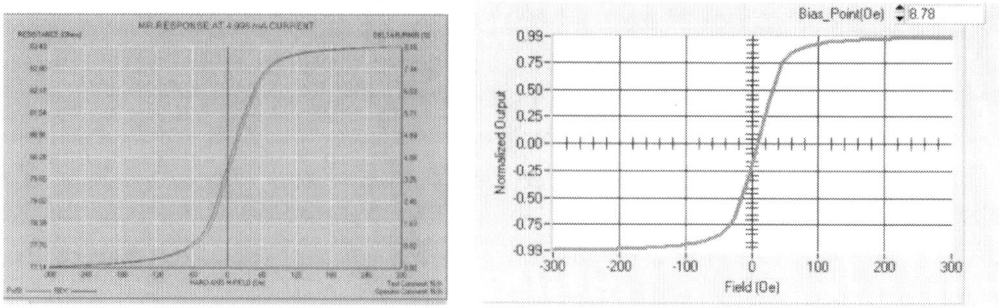

a b

Figure 4 GMR spin-valve transfer curves for D1: a) experimental result, b) normalized simulation result.

Figure 5 shows the GMR transfer curves for Device 2. Although biased with a pair of 400Å CoPt permanents (PM) in the same way, the transducer Device 2 has a smaller dimension than Device 1: the length is decreased to 2.9 um in Device 2 from 14 um in Device 1, the width is reduced to 1.8 um in Device 2 from 2.5 um in Device 1. Again, the experimental and simulated transfer curves are in a very good agreement. Device 2 exhibits an excellent performance showing no hysteresis.

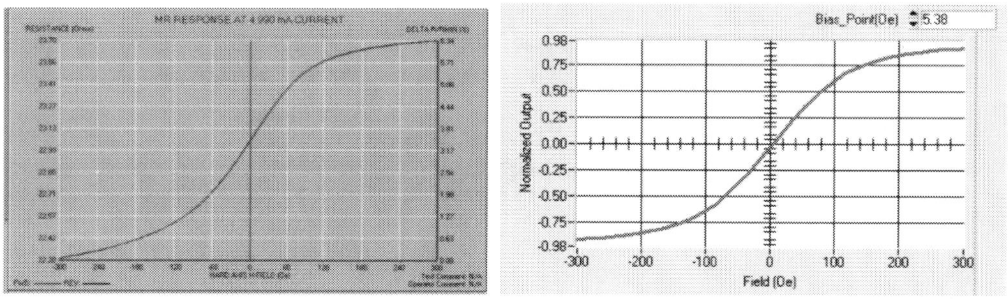

a b

Figure 5 GMR spin-valve transfer curves for D2: a) experimental result, b) normalized simulation result.

From the experimental results, we found that after the device patterning, the GMR is drastically reduced as compared to the initial sheet film level. The GMR spin-valve sensor Device 1 exhibits a reduced GMR of 8.16% from nearly 10% at the initial sheet film level, while sensor Device 2 shows much smaller GMR of 6.34%. The reasons behind this GMR reduction are probably due to two factors: 1) device size effect and 2) series resistance added by sensor/PM junction.

Otherwise, by comparing Figure 4 and Figure 5, it is easy to draw a conclusion that the saturation field of the GMR sensor increases with decreasing the sensor length and width. This can be explained from two aspects: 1) the effective biasing field on the sensing element generated by the permanent magnets increases with decreasing the sensor length; 2) the demagnetization field becomes larger when reducing the sensor width.

4. Conclusion

GMR spin-valve sensors have been fabricated, where the hysteresis is suppressed by placing a pair of CoPt permanent magnets at both ends of the sensing elements. It is found that the size of the spin-valve sensing element has great influence on the device performance. With reduced length and width of the spin valve sensing element, the GMR ratio decreases slightly and the saturation field increases respectively. The fabricated spin-valve sensors exhibited high GMR ratio and good linearity, showing promising applications in linear devices.

Acknowledgement
This work was supported by Innovation Research Team for Spintronic Materials and Device of Zhejiang Province, China.

References

[1] Dieny B, Speriosu V S, Parkin S S P, Gurney B A, Wilhoit D R and Mauri D 1991 *Phys. Rev. B* **43** 1297–1300.
[2] Kanai H, Noma K, Hong J.2001 *Sci. Tech. J.* **37(2)** 174-182.
[3] Qian Z H, Wang D X, Daughton J, Tondra M, Nordman C, Popple A. 2004 *IEEE Tran. Magn.* **40(4)** 2643-2645.
[4] Qian Z H, Wang D X, Daughton J, Lange E, Nordman C, Myers J and Schuetz J 2003 *J. Appl. Phys.* **93** 6870-6872.
[5] Qian Z H, Daughton J, Wang D X, Tondra M 2003 *IEEE Tran. Magn.* **39(5)** 3322-3324.
[6] Heim D E, Fontana R E, Jr., C. Tsang, Speriosu V S, Gurney B A, Williams M L 1994 *IEEE Tran. Magn.* **30(2)** 316-321.
[7] Baubock G, Dang H Q, Hinson D C, Rea L L, Kim Y K 1996 *IEEE Tran. Magn.* **32(1)** 25-30.

Journal of Physics: Conference Series **263** (2011) 012006 doi:10.1088/1742-6596/263/1/012006

The Linewidth Dependence of GMR Properties in Patterned Spin Valve Resistors

Yucheng Sun[1,2], Zhenghong Qian[1*], Ru Bai[1] and Jianguo Zhu[2]

[1]Center For Integrated Spintronic Devices (CISD), Hangzhou Dianzi University, Hangzhou, ZJ 310018, China.
[2]College of Material Sciences and Engineering, Sichuan University, Chengdu, SC 610064, China.

E-mail: zqian@hdu.edu.cn

Abstract. The spin-valve structures with the configuration of NiFeCo(42.5Å)/CoFe(15Å)/Cu(25Å)/CoFe(42.5Å)/CrPtMn(300Å) were fabricated by RF diode sputtering. After a annealing process, the sheet spin-valve thin films were photolithographically patterned into serpentine resistors with different linewidths. The effect of linewidths on the magnetic properties and GMR ratio of patterned spin-valve resistors were studied. It was found GMR decrease with decreasing linewidths due to size effect and the effective anisotropy field H_k and the bias field field H_b are both inversely proportional to linewidths.

1. Introduction

Spin-Valve sensing resistors [1] have attracted much attention and are widely used in hard disks and in a variety of other applications [2-6]. Due to the size effect, patterned Spin-Valve sensing resistors show quite different GMR properties as compared to the sheet Spin-Valve thin films. Exploring the relationship between patterned Spin-Valve sensing resistors' dimension and their GMR properties is very important for optimizing the performance of Spin-Valve based devices. In this work, GMR spin-valves have been patterned into serpentine sensing resistors with different linewidths, the size effect of GMR properties has been investigated with respect to GMR ratio and the effective anisotropy H_k, a mixed effect of shape and uniaxial anisotropies, the offset field H_b.

2. Experiment Procedure

Spin-valve multilayers used in this work were deposited on Si/Si_3N_4(2KÅ) substrates by a RF diode sputtering system. The free and pinned magnetic layers were deposited with a 20Oe field in the same direction. Subsequent to the deposition, the spin-valve multilayers were annealed at 240°C for 2 hours in forming gas with a field of 4000Oe parallel to the easy axes of the magnetic layers in order to lower the free layer dispersion and maximize the pinning strength induced by the pinning layer to the pinned layer. The annealed spin valves show excellent properties, with a pinning field lager than 250Oe. The magnetic properties of the annealed sheet film is summarized in Table 1. Here, H_c represents the coercivity of the free layer, H_{coup} represents the interlayer ('orange peel') coupling between the free and pinned layers..

Table 1 Summary of Magnetic Properties of Sheet Spin-Valves

Hc Oe	H_{coup} Oe	GMR%	Spin-Valve Configuration
3.47	11.5	8.62	**NiFeCo**(42.5Å)/**CoFe**(15Å)/**Cu**(25Å)/**CoFe**(42.5Å)/**CrPtMn**(300Å)

The sheet spin-valve multilayers were then photolithographically patterned into serpentine resistors with different linewidths.

The easy axis of free layer and the pinning direction were patterned along the lengths of patterned spin valves. The pinned layers were then reoriented into the transverse direction by annealing at 240°C for 2 hours in forming gas with a field of 4000Oe applied in the direction perpendicular to the length of patterned spin-valves.

The performance of the spin-valve resistors was measured with a test station, where the magnetic field can be varied from –500 Oe to 500 Oe. During measurement, the field was applied parallel to the sensing axis, which is in the transverse direction of patterned spin-valve sensing resistors. The MR transfer curve was acquired by sweeping the applied magnetic signal field.

3. Results and discussions

3.1. Effect of Patterned Linewidths on the Effective Anisotropy

The effective anisotropy H_k comes from the mixed effect of the shape anisotropy and the uniaxial anisotropy of the free layer. It plays an important role in the determination of sensitivity of patterned spin-valve, and can be expressed as Sensitivity = GMR%/ (2* H_k). In principle, three factors can be used to adjust the H_k. The first one is the orientation of the easy axis of the free FM layer anisotropy. The second one is the patterned spin-valve linewidth. The third one is the thickness of the free FM layer. Figure 1 shows the effect of the patterned spin-valve linewidths on the H_k

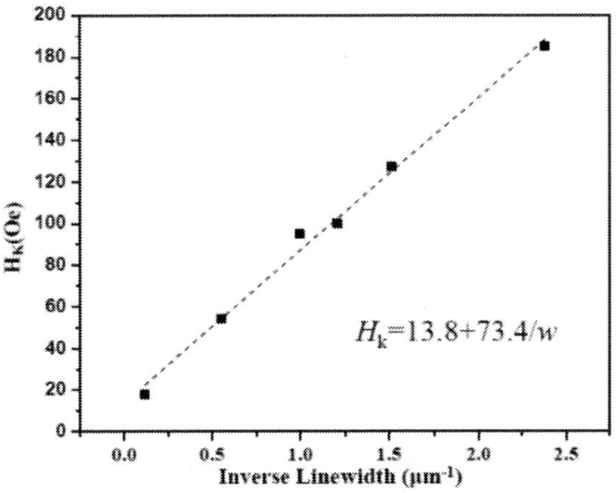

Figure 1. Dependences of H_k on the patterned spin-valve linewidths

It shows that the H_k is inversely proportional to the inverse linewidth of patterned spin-valve resistors, and be expressed by

$$H_k = C_1 + C_2/W \qquad (3.1)$$

Here, The first term C_1 is contributed by the uniaxial anisotropy of the free layer, and the second term C_2/W is contributed by the shape anisotropy of the free layer, where C_2 is constant and W is patterned linewidth. By curve fitting, C_1 and C_2 are determined respectively to be 13.84 Oe and 73.4 Oe•μm for the Wafer 1, C_1 is related to the uniaxial anisotropy field of the free layer.
For NiFeCo layer the uniaxial anisotropy is 21Oe. The discrepancy of the experimental value with the predicted value indicates that the easy axis of the free layer is no longer aligned along the lengths of the patterned spin-valve resistors, it is either randomized or altered away by the annealing process. C_2 is related to both the free layer magnetization M_f and the free FM layer thickness t_f, and can be expressed by a formula of $4\pi M_f t_f$. By calculation, C_2 are 72.2 Oe•μm, and that is very close to the experimental data acquired from the curve fitting.

3.2 Effect of Patterned Linewidths on the Offset Field H_b
The patterned linewidth not only exhibits a strong effect on the effective H_k but also shows a strong effect on the offset field H_b. The origin of the offset is mainly caused by the interlayer 'orange peel' coupling, which is arising from the roughness of the spin-valve multilayer [7]. Note the RKKY related coupling is so small and ignored in this case. The interaction between the free and pinned layer, which is arising from the uncompensated poles at the end of the free and pinned layers, also has an effect on H_b. The third one is the effect of the sensor current, which, in this study, is eliminated by averaging the data taken by passing the current both ways through the patterned spin-valve resistors. Figure 2 shows the dependence of the offset field H_b on the spin-valve linewidth

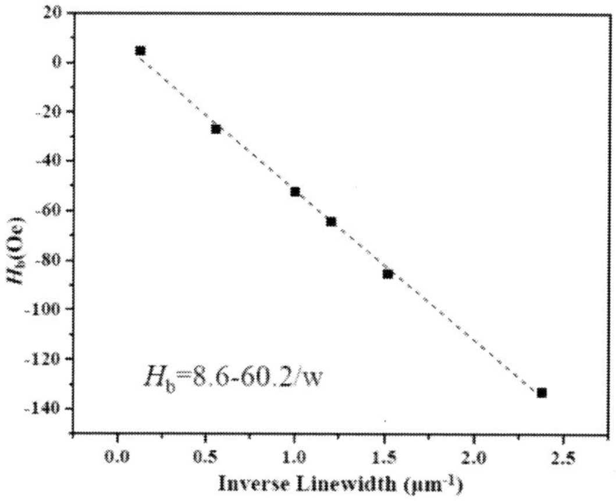

Figure 2. Dependences of H_b on the patterned spin-**valve** linewidths

The offset field H_b appears to be roughly inversely proportional to the linewidth of the spin-valve resistor, and can be expressed by

$$H_b = C_3 - C_4/W \qquad (3.2)$$

Where, the first term C_3 is actually the "orange peel" coupling, and can be determined experimentally by the intercept of the straight line with the offset field axis. The second term $-C_4/W$ is a result of magnetostatic interaction between the free and pinned layers , where C_4 can be determined by the slope of the straight line.
Table 2 summarizes the data from the curve fittings from Fig. 2

Table 2 Data Extricated From Figure 2

C_3(Oe)	C_4 (Oe•μm)
8.6	60.2

In Table 2, it shows that the "orange peel" coupling C_3 determined by the intercept of the straight line with the offset field axis is consistent with the result from the sheet film measurement. The slope C_4 is a measure of the strength of the effect on the free layer generated by uncompensated poles at the edges of the pinned layer, and can be expressed as $C_4 = 4\pi M_p t_p$, where the M_p and t_p are the magnetization and thickness of the pinned layer respectively.

3.3.Effects of Patterned Linewidths on GMR ratio

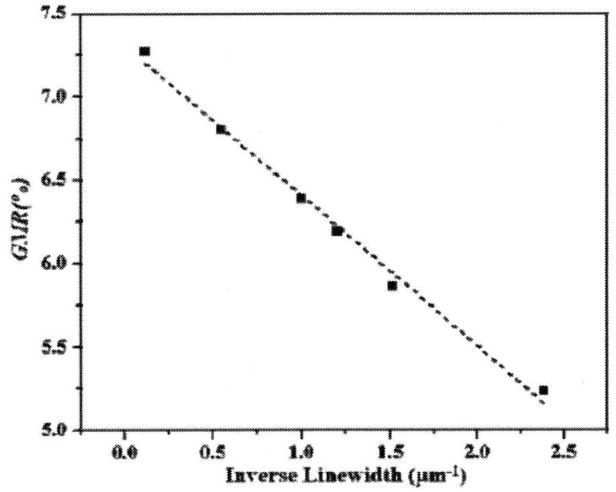

Figure 3. Dependence of GMR ratio on the patterned spin-valve linewidth

As shown in Fig.3, The GMR decreases with decreasing resistor linewidth. The GMR drops from 6.7% with a linewidth of 2 μm to 6.2% with a linewidth of 1μm, and drops further with decreasing linewidth. The effect of the patterned linewidth on the GMR is mainly contributed by two factors, 1) higher probability of electron scattering at edges with narrower linewidth, 2) deterioration of pinning alignment in the transverse direction by stronger demagnetization effect with narrower linewidth.

4. Conclusions

The dependences of magnetic properties on linewidths in patterned spin-valve sensing resistors have been investigated. The GMR ratio is found to decrease with decreasing the linewidths of the patterned spin-valve resistors. Both The effective anisotropy field H_k and the offset field H_b are also size-dependent, and are inversely proportional to the linewidth of patterned spin valve.

Acknowledgement

This work was supported by Innovation Research Team for Spintronic Materials and Device of Zhejiang Province, China.

References

[1] Dieny B, Speriosu V S, Parkin S S P, Gurney B A, Wilhoit D R and Mauri D 1991 *Phys. Rev. B* **43** 1297–1300.

[2] Wolf S A, Awschalom D D, Buhrman R A, Daughton J M, von Molnár S, Roukes M L and Chtchelkanova A Y, Treger D M 2001 *Science* **294** 1488-95.

[3] Qian Z H, Wang D X, Daughton J, Tondra M, Nordman C, Popple A. 2004 *IEEE Tran. Magn.* **40(4)** 2643-2645.

[4] Qian Z H, Wang D X, Daughton J, Lange E, Nordman C, Myers J and Schuetz J 2003 *J. Appl. Phys.* **93** 6870-6872.

[5] Qian Z H, Daughton J, Wang D X, Tondra M 2003 *IEEE Tran. Magn.* **39(5)** 3322-3324.

[6] Daughton J 2003 *Proc. IEEE* **91** 681-6.

[7] Leal J L, Kryder M H 1996 *IEEE. T. Magn* **32** 4642-44.

Journal of Physics: Conference Series **263** (2011) 012007 doi:10.1088/1742-6596/263/1/012007

The Analysis of GMR Sensor's Angular Dependence

Huachen Zhu[1], Zhenghong Qian[1, 2]*, Chunkui Huang[2]

1. Center for Integrated Spintronic Devices, Hangzhou Dianzi University, Hangzhou 310018, China
2. SpinIC, Inc., Yichang 443003, China

*E-mail: zqian@hdu.edu.cn

Abstract. This paper presents the experimental results of GMR sensors' angular dependence, which was tested by applying three different magnetic fields (5Oe, 10Oe, 15Oe) to GMR sensors with different sensitivity. It is found that the GMR sensor is sensitive to angular variation and can be characterized with a cosine function. By comparing the experimental curve with that simulated by the ideal model, we have established a modified model by introducing several new factors to make it fit with the test data.

1. Introduction

Since the giant magnetoresistance (GMR) effect was discovered in 1988 [1, 2], great attention has been paid on its applications, for example as rotation sensor in industry control area. Angular dependence of magnetic sensors, which include Hall-effect device, AMR sensor, and GMR materials have been previously reported [3-5].

However, characterization on angular dependence of GMR sensors is still lacking. In this work, the experiments in characterizing the angular behavior of GMR sensors were carried out in the lab of Center for Integrated Spintronic Devices (CISD) of Hangzhou Dianzi University. Two sensor chips with different sensitivity were chosen and have been measured by applying three discrete external magnetic fields.

The details of the experiment are presented in the section 2 and the experiment results are discussed in section 3. Furthermore, an improved model for angular characterization of GMR sensor is derived by introducing several correction factors in the last section.

2. Experiment procedure

The angular dependence of GMR sensor was measured by applying a fixed field at different direction. These fixed magnetic fields are chosen with three distinctive values of 5Oe, 10Oe and 15Oe. Since the applied fields are high enough so that the geomagnetic field (lower than 0.5Oe which is less than 10% of operation magnetic field) or some other background disturbances can be ignored in this experiment. Two GMR sensor chips with different sensitivity are employed to study the influence of sensor's sensitivity to their angular behaviour. Table 1 shows SA GMR sensor series manufactured by SPINIC INC., in which SA02 and SA03 are chosen for this work. SA02 is with a sensitivity of ~3.5mV/V/Oe, and SA03 is with a sensitivity of ~2.5mV/V/Oe.

Table 1. SA series characteristics [6]

Part Number	Saturation Field (Oe)	Linear Range (|Oe|)		Sensitivity (mV/V-Oe)		Resistance (Ohms)
		Min	*Max*	*Min*	*Max*	
SA00	17.8	0.6	10.2	5.0	6.0	5.5K
SA01	21	0.8	12	4.0	5.1	5.5K
SA02	28	1.0	16	3.0	4.1	5.5K
SA03	39	1.4	20	2.0	3.1	5.5K
SA04	65	2.4	36	1.2	2.1	5.5K

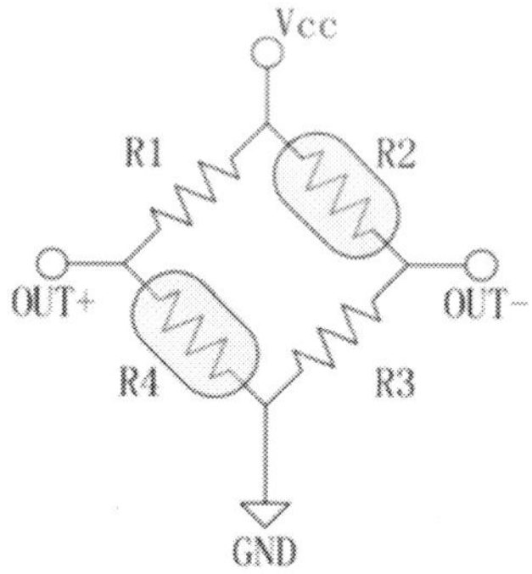

(a)

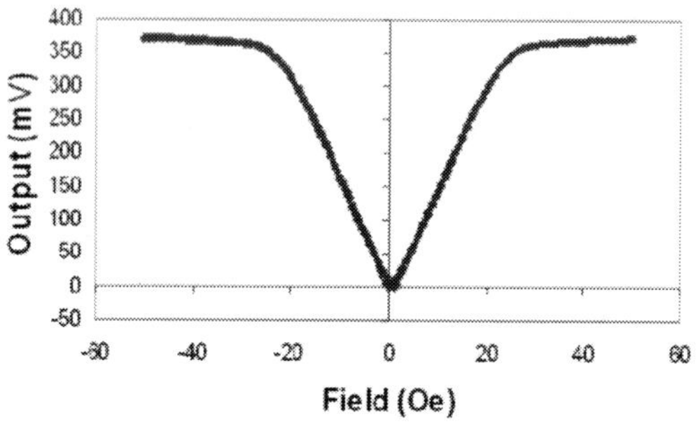

(b)

Figure 2. (a)Wheatstone bridge structure of the sensor, R2 and R4 are covered by magnetic shields. (b)The static characteristics curve of SA02 [7].

Fig. 2 shows the sensor's configuration as well as the transfer curve of SA02 sensor. Its performance is excellent with a good linearity and little hysteresis at near zero fields.

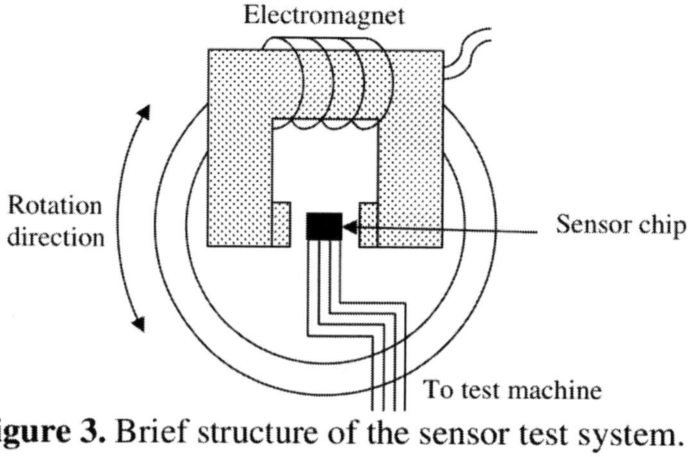

Figure 3. Brief structure of the sensor test system.

The brief structure of test system is displayed in figure 3 composing of GMR sensor chip and electromagnet: the chip is attached on the device-under-test (DUT) board, while the electromagnet used for applying fields can be rotated along the circle flame. The output data was acquired 20 times at each angle and then was averaged.

Because there is no any shield for background magnetic fields, the sensor's sensitivity axis is adjusted to be perpendicular to the earth field, so that the geomagnetic influence could be reduced to the lowest.

3. Results and discussion

The experiment results are presented as following. The sensor SA02's rotation response curves are shown as Figure 4(a), which are obtained respectively applied fields of 5Oe, 10Oe, 15Oe. Fig4 (b) displays the SA03's response results. In these graphs, solid symbol and open symbol indicates the clockwise (CW) and the counter-clockwise (CCW) test results respectively.

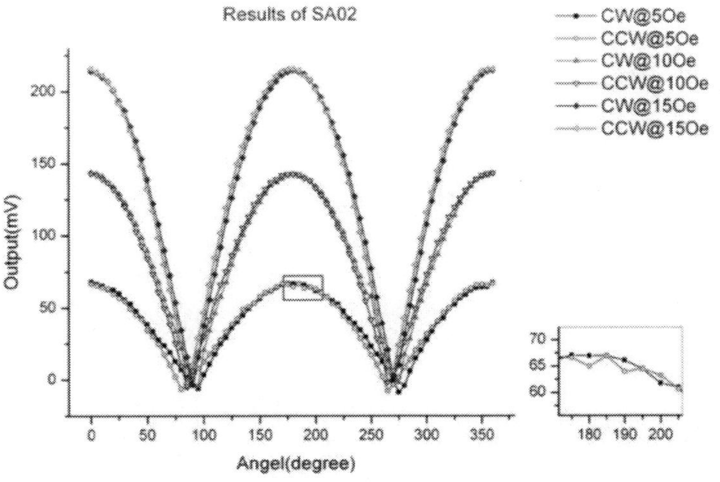

(a)

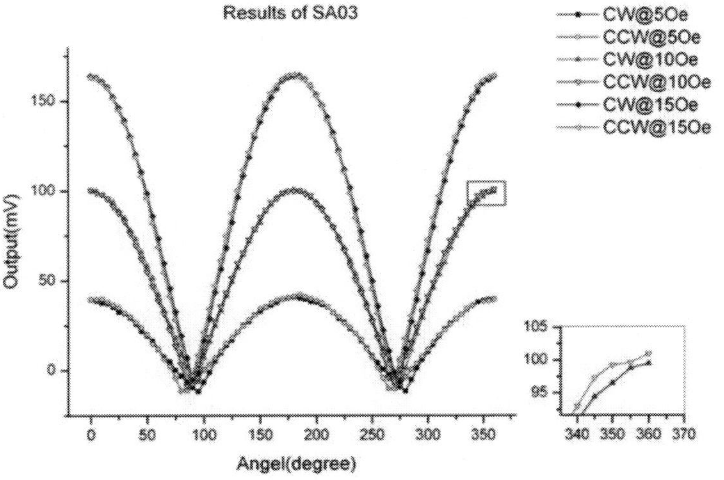

(b)

Figure 4. The rotation response curve. (a) SA02; (b)SA03.

From these curves, we can see the sensor's angular variation can be characterized with a cosine function. Some obvious output vibration appeared nearby the curve peak (shown as in the enlarged charts) at a relatively lower applying field (the 5Oe for SA02, the 5Oe and 10Oe for SA03,) at which magnetic field is roughly parallel to chip's sensitivity-axis. This effect disappears at a higher field. The factors such as the measurement accuracy and the magnetic field distribution generated by electromagnet could cause the error. Nevertheless, after we analyze the chip's sensitivity, we find this fact would be eliminated if the product of chip's sensitivity multiplied by operation field exceeds a certain value (around 25Oe). Therefore, it should be considered in the design for applications.

4. Model and conclusion

4.1. Ideal model

The simple sensing principle of the GMR material can be expressed as Fig. 5. Only the component of magnetic field along the direction of its sensitivity axis can be detected by the sensor.

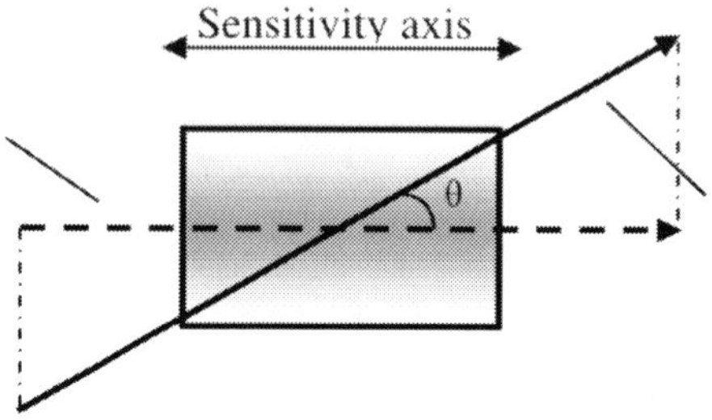

Figure 5. GMR sensor's detecting principle.

The ideal angular characteristic of GMR sensor with Wheatstone bridge structure can be described by equation (1).

$$V_O = V_P \times S\left|H\cos\theta\right| \tag{1}$$

Here V_P is the power voltage, S stands for the sensitivity of a certain chip, while H indicates the magnitude of operation magnetic field.

However, this simple expression is not so accurate if considering some non-ignorable effects in the applications.

4.2. Non-ideal effects

In real application, three non-ideal effects cannot be ignored: 1) the self-offset in chip; 2) the material's sensitivity axis shift; 3) background magnetic noise.

First of all, as expressed in section 2, the sensor is with a configuration of Wheatstone bridge. Any difference among 4-resistor legs could cause the bridge unbalance, and output an offset voltage. Actually, the offset always exists because of process limitation. We use the symbol V_{offset} to indicate the offset voltage.

Secondly, the sensitivity axis shift should be considered. The symbol ω is introduced into the function to express this effect and it is also a constant for a particular sensor.

The third, to the background magnetic noise, only the vector component along the sensitivity axis needs to be considered. It is an external magnetic field bias and variable as position changing, which is marked by H_E.

To sum up these points above, a more practical model should be expressed like equation (2).

$$V_O = V_P \times S\left|H\cos(\theta+\omega)+H_E\right|+V_{offset} \tag{2}$$

Taking the SA02 at 15Oe as example, the-sensitivity is 3mV/V-Oe and is represented by S and ω is calculated out by using experiment's data, which equals to 0.0115π. The V_{offset} and H_E are both taken by actual test, which are -8.74mV and 0.01Oe respectively. Then the model function can be determined as equation (3).

$$V_O = 5\times 3\left|15\cos(\theta + 0.0115\pi) + 0.01\right| - 8.74 \quad (3)$$

Plot the curves of the equation (1) and (3). Fig. 6 displays the comparison result of model vs. data. Obviously, the modified model curve could fit the experimental data better than the ideal model.

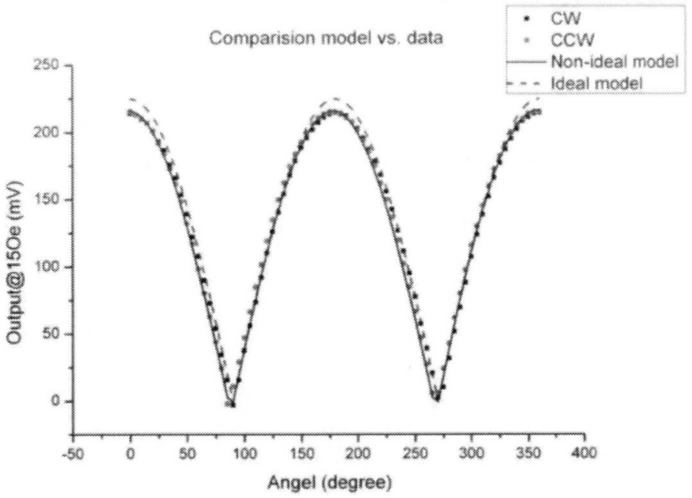

Figure 6. Comparison result of model vs. data.

4.3. Conclusion

By testing the sensor chip SA02 and SA03, and analyzing the experiment results, the angular dependence of GMR sensor is characterized. The influence of sensor's sensitivity on angular behaviour is also analysed. Besides, a modified model is established to describe the characterization of the angular dependence.

Acknowledgement

This work is supported by Innovation Research Team of Spintronic Materials and Device of Zhejiang Province.

References

[1] Baibich M N, Broto J M, Fert A, Van Dau F Nguyen, Petroff F, Etienne P, Creuzet G, Friederich A, Chazelas J, 1988 Giant magnetoresistance of (001)Fe/(001)Cr magnetic superlattices *Phys. Rev. Lett*. **61** 2472

[2] Binasch G, Grünberg P, Saurenbach F, Zinn W, 1989 Enhanced magnetoresistance in layered magnetic structures with antiferromagnetic interlayer exchange *Phys. Rev. B*. **39** 4828

[3] Jen S U, Huang K P, Lee J C, 1998 The angular dependence of the planar Hall effect in Co-Ni films *J. Appl. Phys.* **84** 843

[4] Vedyayev A, Dieny B, Ryzhanova N, Genin J B, Cowache C, 1994 Angular Dependence of Giant Magnetoresistance in Magnetic Multilayered Structures *Europhys. Lett.* **25** 465

[5] Song J S, Lee W J, Min B K, and Chung G S, 2001 Wheastone-bridge-type MR sensors of Si(001)/NiO(300 [A-ring])/NiFe bilayer system *Proc. SPIE* **4601** 370

[6] Datasheet—SA series GMR sensor. SpinIC, Inc. http://www.spinic.com.cn/

[7] Qian Z, Bai R, Huang C, Wu J, 2009 Advanced Spintronic Materials and Devices *Instrument Technique and Sensor* **z1** 96

Dependence of $BiFeO_3$ thickness on exchange bias in $BiFeO_3$/ Co_2FeAl multiferroic structures

X Zhang, D L Zhang, Y H Wang, J Miao*, X G Xu, Y Jiang*

State Key Laboratory for Advanced Metals and Materials, School of Materials Science and Engineering, University of Science and Technology Beijing, Beijing, 100083, China

*Email: j.miao@ustb.edu.cn, yjiang@ustb.edu.cn

Abstract. We have grown $BiFeO_3$ (BFO) thin films with different thickness on $Si/SiO_2/Ti/Pt(111)$ substrates by pulsed laser deposition. Half-metallic Co_2FeAl (CFA) films with a thickness of 5 nm were then grown on the BFO films by magnetron sputtering. Through the magnetic hysteresis loops of the BFO/CFA heterostructure, we observe a direct correlation between the thickness of the BFO film and exchange bias (EB) field. The EB field exhibits fluctuation behavior with a cyclical BFO thickness of 60 nm, which is close to the spiral modulation wavelength (62 nm) of BFO. It indicates the influence of spiral modulation on the EB in the BFO/CFA multiferroic structure.

1. Introduction

Multiferroics combining ferromagnetic and ferroelectric are a time-honored subject of study and have led to some of the most important technological advances to date [1]. One of the most widely studied multiferroic materials is $BiFeO_3$ (BFO), which possesses a rhombohedrally distorted perovskite structure at room temperature [2]. BFO has ferroelectric Curie and magnetic Néel temperatures at ~1100K and 647K, respectively [3-6], which indicates their potential applications in date storage, spintronic sensors, and so on [7, 8]. BFO has an antiferromagnetic spin ordering, but it also displays a weak ferromagnetic moment arising from a canted spin structure [9, 10]. Followed Wang et al., who have grown epitaxial BFO films on $SrTiO_3(100)$ substrates [11], a great many studies have been done to grow high quality, fully epitaxial, and single phase thin films of BFO on various substrates [12-15].

Recently, electric-field control of magnetic exchange, and even spin polarization [16], has attracted much attention. Multiferroic materials, such as $YMnO_3$ and BFO, have been grown as the bottom antiferromagnetic layer in spin-valves [17], or the barrier layer in magnetic tunnel junctions [18]. In the devices, the magnetoelectric coupling was designed to tune the effective direction of the exchange bias (EB) by an electric field. Dynamic switching of the EB field with an applied electric field has been reported by Laukhin et al. focusing on $YMnO_3$ at very low temperatures [19]. Martin et al. [20] studied the asymmetric spin valves with BFO as the antiferromagnetic layer. However, till now there is few study on the influence of the BFO thickness on the EB.

In this letter, we study the effect of BFO thickness on the EB field in a BFO/Co_2FeAl (CFA) heterostructure and find that the EB field exhibits a fluctuation behavior with a cyclical BFO thickness of 60 nm, which is close to the spiral modulation wavelength (62 nm) of BFO. It indicates the influence of spiral modulation on the EB in the BFO/CFA multiferroic structure.

2. Experiment

The BFO films with different thickness were deposited on $Si/SiO_2/Ti/Pt(111)$ substrates by pulsed laser deposition. The target was prepared by mixing Bi_2O_3 and Fe_2O_3 in a 1.2:1 stoichiometric ratio then sintering at 820 °C in air. The CFA films with a thickness of 5 nm were grown on the BFO films by magnetron sputtering. A 700 Oe local magnetic field was applied in order to induce an easy axis during the sputtering. 4 nm Ta films were then grown on the top of the heterostructure to protect the CFA films from oxidation. The structural characterization was performed using a M21XVHF22 X-ray diffractometer with Cu/Kα radiation. The surface morphology and cross section images were obtained using Scanning Electron Microscopy (SEM). Magnetic hysteresis loops were measured using an Alternating Gradient Magnetometer (AGM).

3. Results and discussion

As shown in figure 1(a), the X-ray diffraction (XRD) pattern of the BFO/Pt(111) structure demonstrates a high quality and single phase BFO film. An enhancement in (111) orientation is clearly observed due to the bottom electrode Pt(111) layer. According to the formula $\alpha_{(111)} = I(111)/[I(100)+I(110)+I(111)]$, the calculated degree of (111) orientation is above 50%. Indeed, several different orientations have been reported for BFO thin films deposited by different techniques and on different buffers and substrates. It is proved that the (111)-oriented BFO thin films generally exhibit better ferroelectric behavior than those of other orientations [21-25]. Figure 1 (b) and (c) are the surface morphology and cross section of the BFO film. The BFO film shows a dense surface with grains which could be classified into two kinds of diameters of 30 nm and 100 nm. It can be seen from figure 1(c) that the grains were grown both perpendicular (left arrow) and parallel (right arrow) to the substrate surface. These results indicate the existence of differently oriented grains on the film surface. Lahmar et al. [26] also have got similar results. The parallel grains were considered to be formed as the result of local interaction between the film and the Pt-bottom electrode.

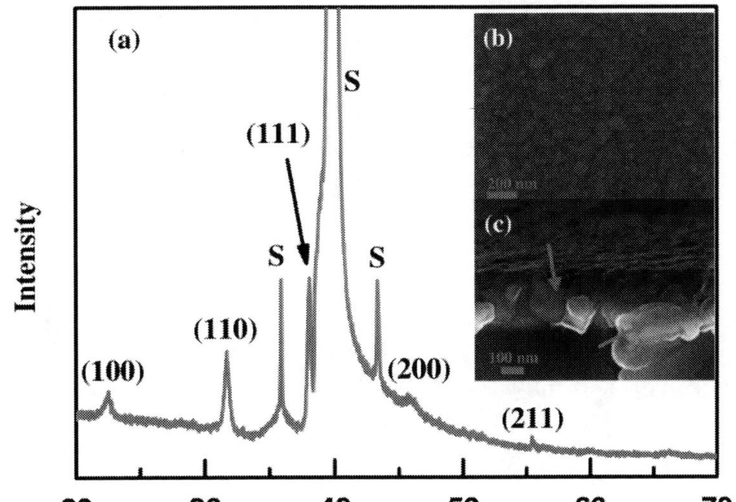

Figure 1. Structural analysis of the BFO films deposited on a $Si/SiO_2/Ti/Pt$ (111) substrate. (a) XRD pattern; (b) SEM of the surface morphology; (c) SEM of the cross section.

We have fabricated the BFO/CAF heterostructures of $Si/SiO_2/Ti/Pt(111)$/BFO (t nm)/CFA (5 nm). Here t=100, 120, 140, 160, 180 and 200 nm, respectively. Figure 2 gives the magnetic hysteresis loops for these heterostructures measured at room temperature. The hysteresis loops are enlarged and shifted along the field axis compared to that of the CFA monolayer, which is shown in figure 3 (a), demonstrating the presence of EB at the CFA/BFO interfaces. In figure 3 (b), when H is applied perpendicular to the films, the M(H) shows a typical hard axis behavior.

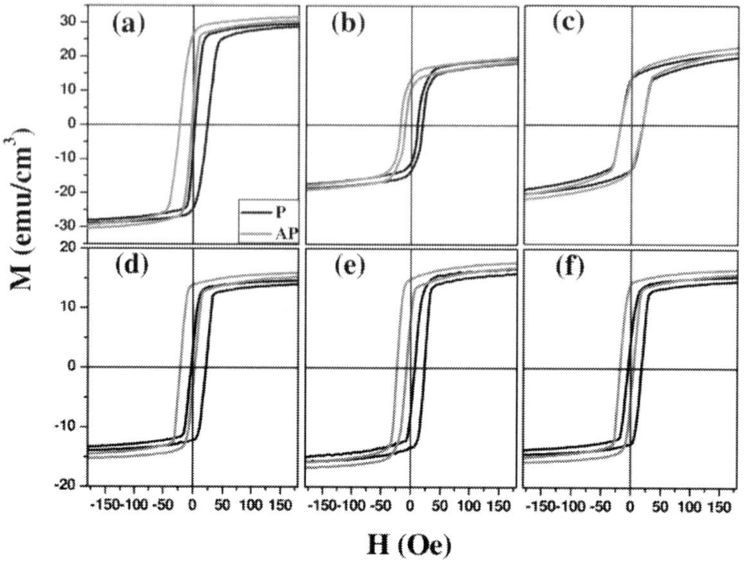

Figure 2. Room temperature M-H loops of BFO/CFA structures with different BFO thickness of (a) 100 nm, (b) 120 nm, (c) 140 nm, (d) 160 nm, (e) 180 nm, and (f) 200 nm, respectively.

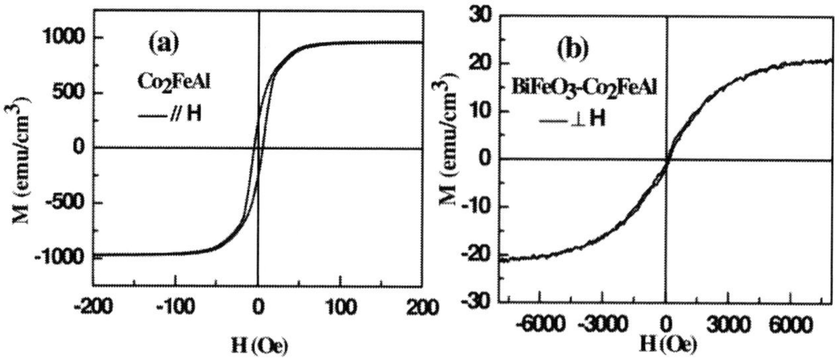

Figure 3. Room temperature M-H loops of (a) CFA monolayer and (b) along perpendicular direction of BFO/CFA films.

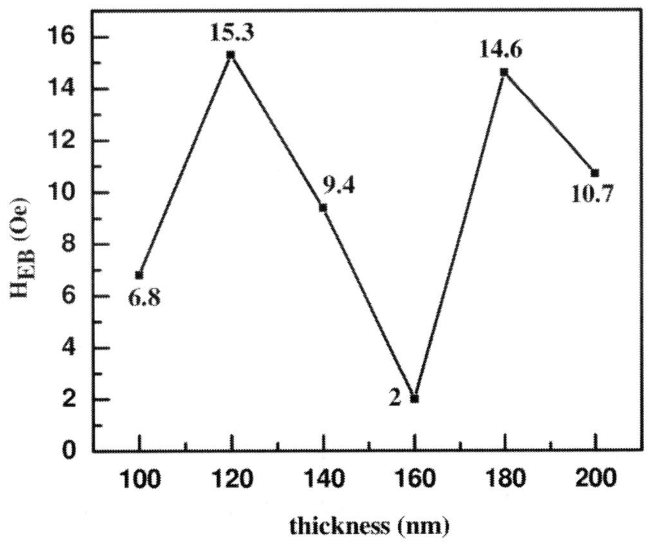

Figure 4. The H_{EB} vs. the BFO thickness of the BFO/CFA structures.

Figure 4 shows the correlation between the EB field (H_{EB}) and the BFO thickness t. The values of H_{EB} are 6.8, 15.3, 9.4, 2, 14.6, and 10.7 Oe, for t= 100, 120, 140, 160, 180 and 200 nm, respectively. The H_{EB} exhibits a fluctuation behavior with a cyclical thickness of 60 nm, which is close to the spiral modulation wavelength (62 nm) of BFO. The results are quite different from the work of Béa [18] and Nogués [27]. There are two reasons may explain the fluctuation of H_{EB}. Firstly, the different orientation of BFO, as shown in figure 1(c), could distract the interface effect. Nogués has also reported that the orientation of AFM layers are no longer stable above certain thickness, and this microstructual change leads to a decrease in H_{EB}. The lower value of H_{EB} may also due to this reason. Secondly, the effect of the BFO spiral modulation structure has been unveiled when H_{EB} has leveled off, which is concealed when H_{EB} is changing abruptly. In other words, the BFO spiral modulation structure could affect H_{EB}, in a limited range.

4. Conclusion

In summary, we have grown the BFO/CFA heterostructures on $Si/SiO_2/Ti/Pt(111)$ substrates and observed distinct EB behavior at room temperature. The H_{EB} exhibits a fluctuation behavior with a periodic thickness of 60 nm, which may be caused by various orientation and the spiral modulation structure of the BFO films. The next work would be growing single oriented BFO film to demonstrate the correlation between the H_{EB} and the BFO thickness.

ACKNOWLEDGMENTS

This work was partially supported by the National Natural Science Foundation of China (Grant Nos. 50701005, 50831002, 50971025), the National Basic Research Program of China (Grant No. 2007CB936202) and the Key Grant Project of Chinese Ministry of Education (No. 309006).

References

[1] Prellier W, Singh M P and Mulugavel P 2005 *J. Phys.: Condens. Matter* **17** R804
[2] Schmidt R, Eerenstein W, Winiecki T, Morrison F D and Midgley P 2007 *Phys. Rev. B* **75** 245111

[3] Yun K Y, Ricinschi D, Kanashima T and Okuyama M 2006 *Appl. Phys. Lett.* **89** 192902
[4] Kim D H, Lee H N, Biegalski M D and Christen H M 2008 *Appl. Phys. Lett.* **92** 012911
[5] Ranjith R, Kundys B and Prellier W 2007 *Appl. Phys. Lett.* **91** 222904
[6] Nakamura Y, Nakashima S, Ricinschi D and Okuyama M 2008 *Functional Materials Letters* **1** 19
[7] Yuan G L, Or Siu Wing and Chan H L W 2007 *J. Appl. Phys.* **101** 024106
[8] Wang H, Zheng Y, Cai M Q, Huang H T and Chan H L W 2009 *Solid State Commun.* **149** P641
[9] Sosnovska I, Peterlin-Neumaier T and Steichele E 1982 *J. Phys. C* **15** 4835
[10] Blaauw C and van der Woude F 1973 *J. Phys. C: Solid State Phys.* **6** 1422
[11] Wang J, Neaton J B, Zheng H, Nagarajan V, Ogale S B, Liu B, Viehland D, Vaithyanathan V, Schlom D G, Waghmare U V, Spaldin N A, Rabe K M, Wuttig M and Ramesh R 2003 *Science* **299** 1719
[12] Béa H, Bibes M, Zhu X -H, Fusil S, Bouzehouane K, Petit S, Kreisel J and Barthélémy A 2008 *Appl. Phys. Lett.* **93** 072901
[13] Chen Y C, Lin Q R and Chu Y H 2009 *Appl. Phys. Lett.* **94** 122908
[14] Cheng Z X, Wang X L, Kimura H, Ozawa K and Dou S X 2008 *Appl. Phys. Lett.* **92** 092902
[15] Tian W, Vaithyanathan V and Schlom D G 2007 *Appl. Phys. Lett.* **90** 172908
[16] Garcia V, Bibes M, Bocher L, Valencia S, Kronast F, Crassous A, Moya X, Enouz-Vedrenne S, Gloter A, Imhoff D, Deranlot C, Mathur N D, Fusil S, Bouzehouane K and Barthélémy A 2010 *Science* **327** 1106
[17] Dho J, Qi X D, Kim H, Macmanus-Driscoll J L and Blamire M G 2006 *Adv. Mater.* **18** 1445
[18] Béa H, Fusil S, Bibes M, Cherifi S, Locatelli A, Warot-Fonrose B, Herranz G, Deranlot C, Jacquet E, Bouzehouane K and Barthélémy A 2006 *Appl. Phys. Lett.* **89** 242114
[19] Laukhin V, Skumryev V, MartíX, Hrabovsky D, Sánchez F, García-Cuenca M V, Ferrater C, Varela M, Lünders U, Bobo J F and Fontcuberta J 2006 *Phys. Rev. Lett.* **97** 227201
[20] Martin L W, Chu Y H, Zhan Q and Ramesh R 2007 *Appl. Phys. Lett.* **91** 172513
[21] Singh S K, Kim Y K, Funakubo H and Ishiwara H 2006 *Appl. Phys. Lett.* **88** 162904
[22] Bai F M, Wang J, Wuttig M, Li J F, Wang N, Pyatakov A P, Zvezdin A K, Cross L E and Viehland D 2005 *Appl. Phys. Lett.* **86** 032511
[23] Li J F, Wang J L, Wuttig M, Ramesh R, Wang N, Ruette B, Pyatakov A P, Zvezdin A K and Viehland D 2004 *Appl. Phys. Lett.* **84** 5261
[24] Wu J G and Wang J 2009 *J. Appl. Phys.* **106** 054115
[25] Wu J G and Wang J 2009 *J. Appl. Phys.* **106** 104111
[26] Lahmar A, Habouti S, Solterbeck C H, Dietze M and Es-Souni M 2010 *J. Appl. Phys.* **107** 024104
[27] Nogués J and Scholler I K 1999 *J. Magn. Magn. Mater.* **192** 203

Journal of Physics: Conference Series **263** (2011) 012009 doi:10.1088/1742-6596/263/1/012009

Delta-Sigma Digital Current Sensor Based On GMR

Zhili Wang, Zhenghong Qian*, Xiaowei Huang

Center for Integrated Spintronic Devices, Hangzhou Dianzi University, Hangzhou 310018, China.

*Email: zqian@hdu.edu.cn

Abstract. This paper presents an integrated delta-sigma digital current sensor. With high sensitivity and good linearity, GMR sensor is a good choice for current measurement application. In this design, a GMR sensor which is configured as a Wheatstone bridge of four spin-valve sensing elements is integrated with CMOS signal processing circuit together, and the output signal is converted to digital domain by an on-chip 10-bit delta-sigma ADC. The whole circuit is designed and simulated using CSMC standard 0.5 μm process.

1. Introduction

Current sensors are widely used as feedback devices in field industrial equipment, power system protection and various test-controlled systems. Defined by the working principle, current sensors could be classified as traditional current transformer, current shunt, Rogowski current sensor, fluxgate current sensor and Hall-effect current sensor, AMR current sensor and GMR current sensor etc. [1].

Nowadays, the most commonly used current sensor is hall-effect current sensor, which is usually used to measure large current. However, due to the low sensitivity and poor temperature stability of hall-effect sensors [2], it could not meet the requirement for precise measurement, especially for low current measurement.

These shortcomings could be overcome by use GMR sensor. Single-module integrated GMR current sensor is tiny. In order to fit for small size sensors, a delta-sigma AD converter using over sampling technology is employed, which provides high resolution and excellent linearity without using precise or trimmed components [3].

In this work, a GMR sensor and a second-order delta-sigma modulator are integrated together. Standard 0.5 μm process is adopted, with the whole system operating under ±2.5 V supply voltage. The simulation shows the sensor could achieve a resolution of 0.5 mA in the operation range of ±80 mA.

2. Integrated digital GMR current sensor design

2.1. Structure and principle

The linear spin-valve sensor [4] is very suitable for small current sensing as it has desired high sensitivity and good linearity. The sensor is with an operation magnetic range of +15 Oe to -15 Oe, no visible hysteresis and the sensitivity is 0.094%/Oe.The designed digital current sensor is shown in Figure1.

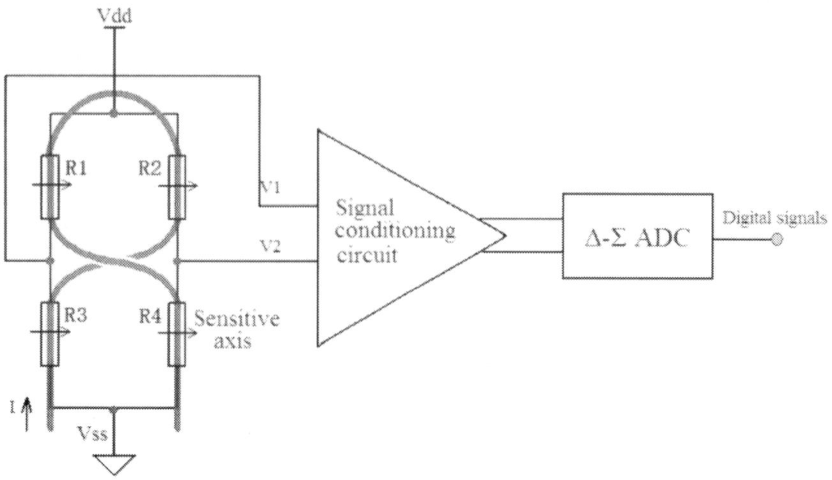

Figure 1. Architecture of current sensor

Shown in Figure 1 are 4 identical spin-valve resistors making up a Wheatstone bridge. The advantage of this configuration is providing temperature compensation. A current wire is placed along with the resistors of the Wheatstone bridge, and the output of the bridge is conditioned and magnified by the conditioning circuit, and then the signal is converted to digital by a 10-bit delta-sigma ADC. According to Ampere Circuit Theorem, when electricity current flowing through an infinitely long wire, the magnetic field strength H induces at the distance r from the wire is $H = I/2\pi r$. Therefore, if the sensors bridge is put at the place 100 μm to the current wire, a -10 Oe to +10 Oe magnetic field will be induced when the current I varies from -80 mA to +80 mA , and the output of the bridge can reflect the current perfectly.

2.2. Signal conditioning circuit

In order to make the sensor sensitive and accurate, following the Wheatstone bridge a signal conditioning circuit is designed.

The output of the Wheatstone bridge is

$$\Delta V = V1 - V2 = \frac{Vdd - Vss}{R1(1-\Delta) + R3(1+\Delta)} \cdot R3(1+\Delta) - \frac{Vdd - Vss}{R2(1+\Delta) + R4(1-\Delta)} \cdot R4(1-\Delta) = (Vdd - Vss) \cdot \Delta$$

here Δ is the change rate of the resistors, and it will be zero in ideal case with no current input. In fact, since it is difficult to match the GMR resistors perfectly in the fabrication, plus the interference of ambient magnetic field, output offset is inevitable. It is the main factor that influences the precision of the GMR sensor. Meanwhile, according to the study in [4], the bridge sensor could reach almost 50 mV signal output in the magnetic range of ±10 Oe. In order to convert it into digital signal accurately, a conditioning circuit is used to adjust the bridge sensor signal. From the above discussion, here a two-op-amp instrumentation amplifier and another amplifier are incorporated as the conditioning circuit to compensate the offset [5].

The conditioning circuit is shown in Figure 2. A1, A2 and A3 are three identical amplifiers. A1 and A2 constitute a two-op-amp instrumentation amplifier. V1, V2 are the outputs of Wheatstone bridge, hence

$$Vout1 = (V1 - V2)\left(1 + \frac{R2}{R1} + \frac{2 \cdot R2}{Rg}\right) + Vref0 \text{ when } R1 = R3,\ R2 = R4.$$

With the amplifier A3, Vout2 is opposite with Vou1 referenced to ground when $R5 = R6$.

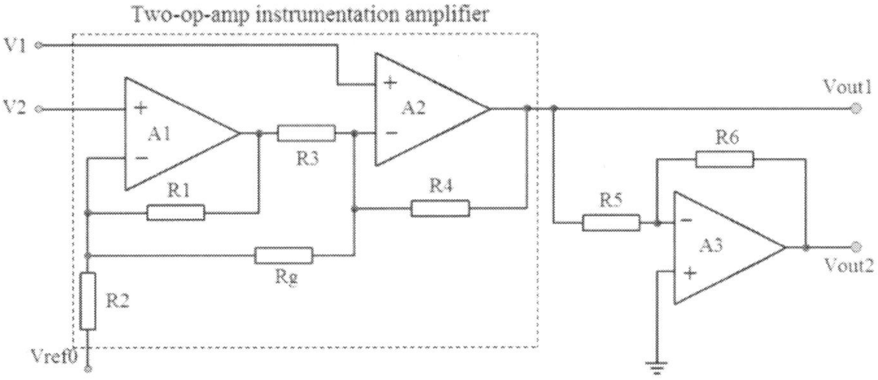

Figure 2. Schematic of signal conditioning circuit

Here, Rg is an external resistor employed to adjust the gain of the circuit. And Vref0 is the external reference voltage which is used to compensate the offset of the bridge and keep the output of the sensor zero when there is no signal input.

It requires high stability and low power consumption in the application of the current sensor, hence, a simple two-stage OTA (Operational Transconductance Amplifier) is chosen to realize the modules A1, A2 and A3 in Figure 2.

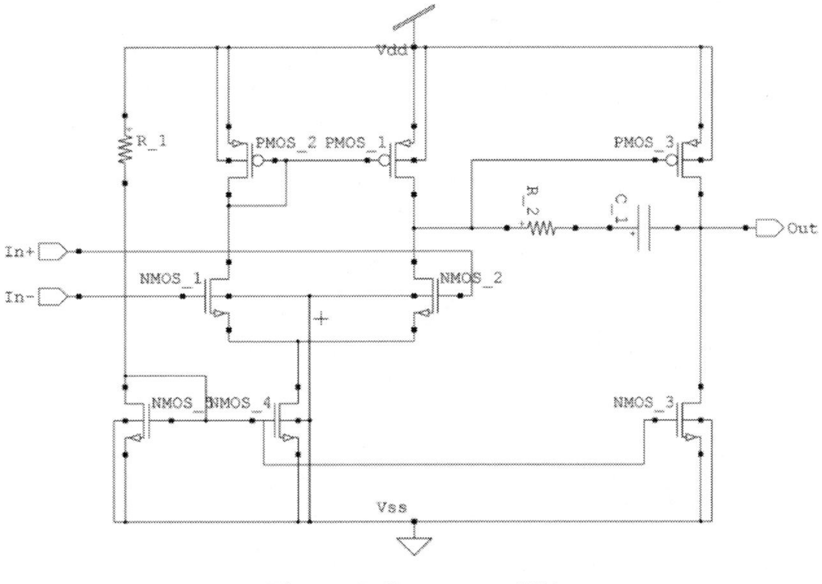

Figure 3. Two-stage OTA

The two-stage OTA is shown in Figure 3. The circuit operates with ±2.5 V voltage supply. The pins labeled “In+” and “In-” correspond to the positive input and negative input respectively. Resistor R_2 and capacitor C_1 are used to optimize the phase margin of the OTA. In order to insure the performance of the circuit, the gain of the OTA should be more than 70 dB, and phase margin more than 60° .

Figure 4 shows the simulation result of the two-stage OTA with CSMC standard 0.5 μm process. From it we can see the gain is 78dB, phase margin is 65° and bandwidth is more than 100MHz, which are more than satisfactory.

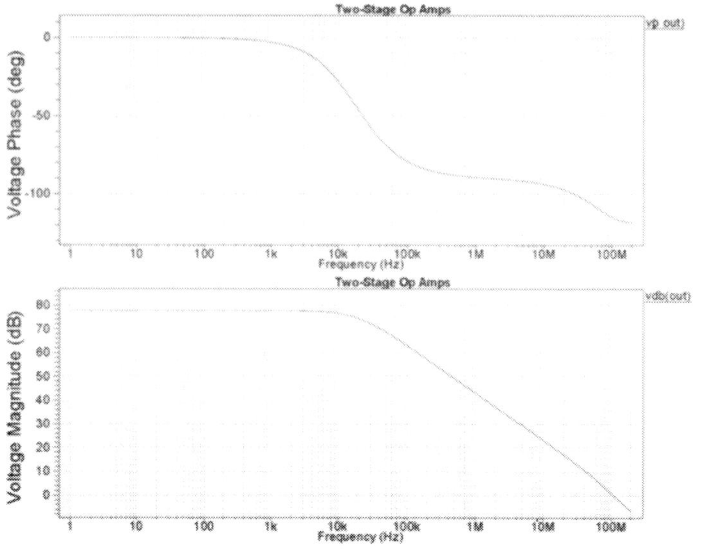

Figure 4. Gain and phase-margin of the two-stage OTA

2.3. Second-order delta-sigma AD converter

Delta-sigma AD converter has simple structure and high accuracy, so it is very suitable for using in single-module sensors.

The block diagram of a second-order delta-sigma AD converter is shown in Figure 5. It consists of two parts, one is delta-sigma modulator, and the other is digital filter. Modulator is the hard-core of the delta-sigma AD converter. It consists of two integrators, a 1bit ADC (a comparator) and a 1 bit DAC (switches). The closed loop forces the value of the feedback signal to track the input value of the modulator. Any difference between feedback and input is accumulated in the integrator until the flip of the ADC output that makes the sign of feedback-input change [3] [6].

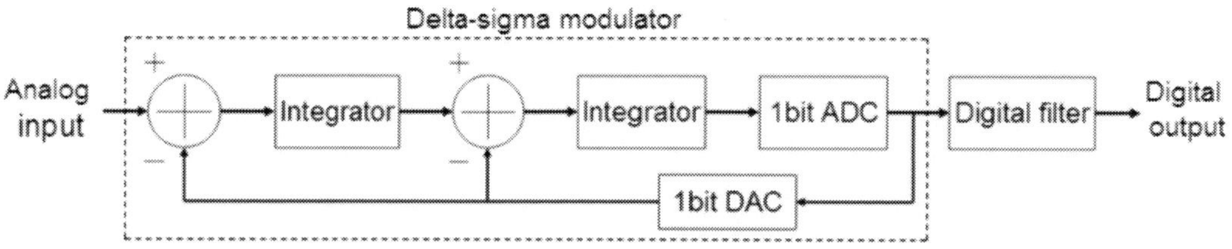

Figure 5. Second-order delta-sigma AD converter

In the current sensor, the expected operation range is ±80 mA and the expected resolution is 0.5 mA, $2^8 < \frac{160}{0.5} < 2^9$, so a delta-sigma modulator with 9-bit ENOB (Effective Number of Bits) is enough. However, considering there will be some errors, in order to insure the resolution and linearity a 10-bit one is chosen [7] [8].

Journal of Physics: Conference Series **263** (2011) 012009 doi:10.1088/1742-6596/263/1/012009

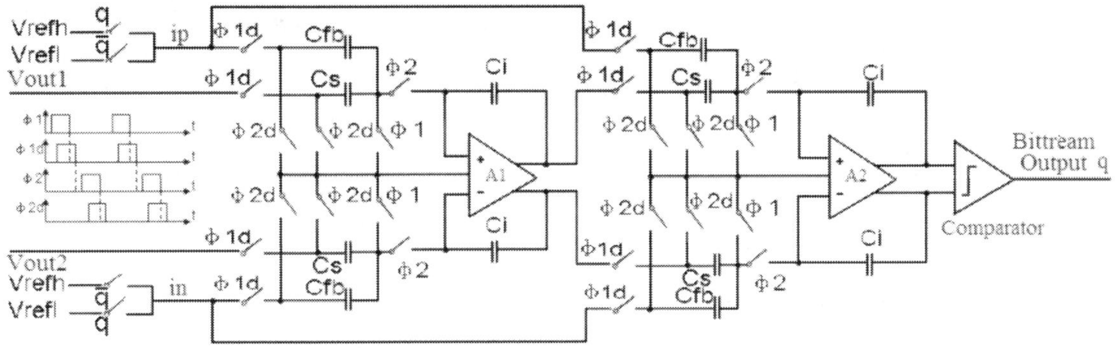

Figure 6. Schematic of the second-order delta-sigma modulator

The circuit schematic of the second-order delta-sigma modulator is shown in Figure 6. Two fully differential folded-cascode amplifiers are used to realize the modules A1 and A2, and a comparator with latch follows the second integrator. In order to avoid the clock feed-through of the switched-capacitor integrator[9], Φ 1 and Φ 2, Φ 1d and Φ 2d are two-phase non-overlapping clocks, Φ 1d, Φ 2d are the signal Φ 1, Φ 2 with delay respectively. The q is the output of the comparator. When q is high, Vrefh is connected to ip, Vrefl is connected to in, and when q is low, Vrefl is connected to ip, Vrefh is connected to in. Here, Vrefh is 1.5 V and Vrefl is -1.5 V. The sampling capacitor Cs and Cfb both are 1 pf, integrating capacitor Ci is 2 pf.

3. Simulation results

The delta-sigma modulator is simulated based on MATLAB with the output of the instrumentation amplifier as input. Figure 7 shows the model of second-order white noise delta-sigma modulator based on MATLAB SIMULINK. The model includes many non-ideal effects, such as clock jitter, switch nonlinearity, kT/C noise and amplifier noise etc.

When a 50 Hz, amplitude of 80 mA AC current flowing through the current wire, according to the sensitivity of the spin-valve resistor, the output of the Wheatstone bridge is

$$\Delta Vout_{max} = (Vdd - Vss)\cdot \Delta \cdot H_{max} = 5V \times 0.094\% / Oe \times 10Oe = 47mV,$$

here Δ is the change rate of the resistors. So it would be a sine curve with the same frequency and amplitude of 47 mV. By adjusting the external resistor Rg and external reference voltage Vref, the output of the instrumentation amplifier is in the range ± 2 V with no offset.

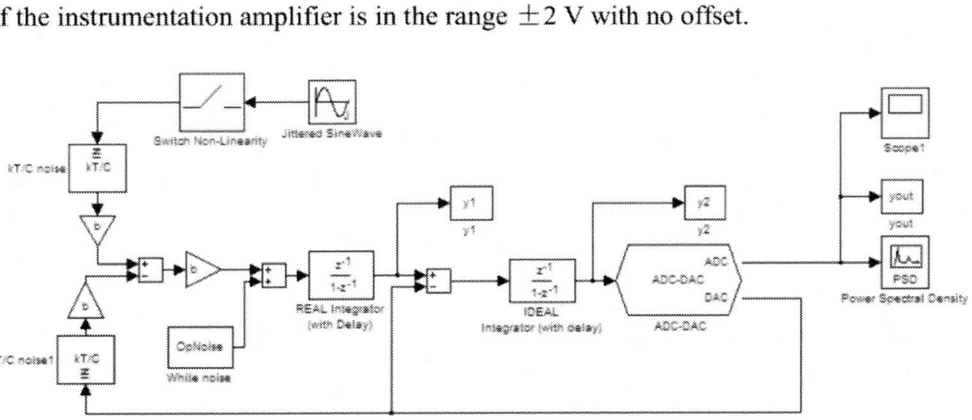

Figure 7. Model of second-order white noise delta-sigma modulator

With the OSR (Over Sampling Ratio) of 128, sampling frequency of 12.5 KHz, the simulation results of the delta-sigma modulator input and output are displayed in Figure 8. And the power spectral density of the modulator is shown in Figure 9, from which we can see the SNDR of the modulator is 65.0 dB, i.e. ENOB is 10.49 bits.

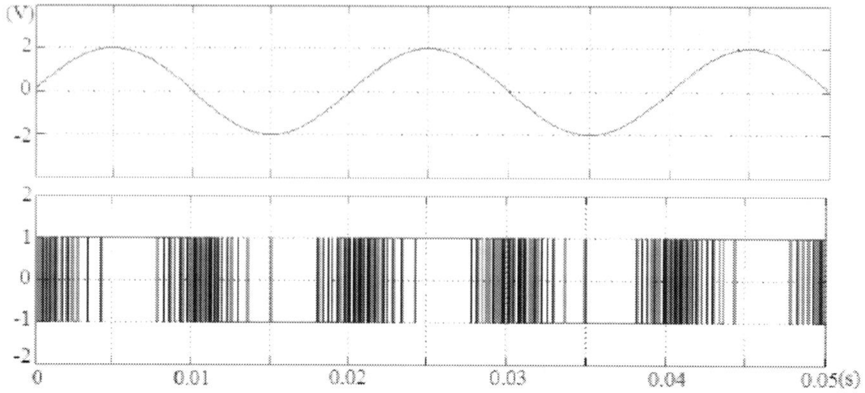

Figure 8. Transient waveforms of Δ-Σ modulator input and output

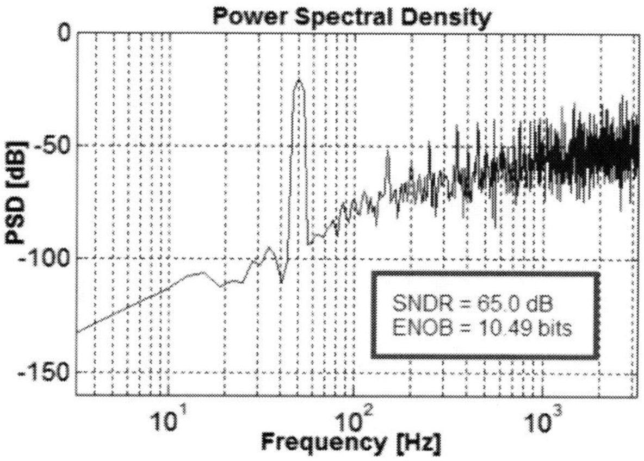

Figure 9. PSD of the delta-sigma modulator

4. Conclusions

A delta-sigma digital current sensor with GMR sensing element integrated with an AD converter has been designed and simulated. In this work, the un-negligible bridge offset is compensated with an external voltage, and then the output signal is converted to digital with an on-chip 10-bit delta-sigma ADC. Simulation shows that the power dissipation of the OTA is 369 μ W, the calculated power of the Wheatstone bridge is 5 mW, so the whole dissipation of the sensor should be less than 100 mW. And the sensor would have the resolution of 0.5 mA which is expected.

ACKNOWLEDGMENT

This work is supported by Innovation Research Team of Spintronic Materials and Device of Zhejiang Province.

References

[1] P.Ripka 2004 Current Sensors Using Magnetic Materials *Journal of Optoelectronics and Advanced Materials* **Vol. 6 No. 2** 587 – 592

[2] Qian Zhenghong, Bai Ru, Huang Chunkui and Wu Jiande 2009 Advanced Spintronic Materials and Devices *Instrument Technique and Sensor* **supplement** 96-101

[3] Jouini M, Guessab S and Kielbasa R 2001 High-level design of a digital current sensor using multibit sigma-delta modulation *Sensor for Industry, 2001. Proceedings of the First ISA/IEEE Conference* **vol. no.** 23-28

[4] Qian Z., Wang, D., Daughton J. M., Tondra M. and Nordman C. Linear Spin-Valve Bridge Sensing Devices *A.IEEE Transactions on Magnetics* **vol. 40 issue 4** 2643-2645

[5] Walt Jung 2004 *Op Amp Applications Handbook* (Beijing: POSTS & TELECOM PRESS) p 92

[6] de Mori C.R.T., Crepaldi P.C. and Pimenta T.C. 2001 A 3-V 12-bit second order sigma-delta modulator design in 0.8-μm CMOS *Integrated Circuits and Systems Design 14th Symposium on.* **vol. no.** 124-129

[7] Beydoun A., Jabbour, C., Fakhoury, H., Nguyen, V.-T., Naviner, L. and Loumeau, P. 2009 A 65 nm CMOS versatile ADC using time interleaving and ΣΔ modulation for multi-mode receiver *Circuits and Systems and TAISA Conference Joint IEEE North-East Workshop on* **vol. no.** 1-4

[8] Richard Schreier and Gabor C. Temes 2005 *Understanding delta-sigma Data Converters* (NJ: John Wiley & Sons Inc)

[9] Philip E. Allen and Douglas R. Holberg 2005 *CMOS Analog Circuit Design* (Beijing: Publishing house of electronics industry)

Journal of Physics: Conference Series **263** (2011) 012010 doi:10.1088/1742-6596/263/1/012010

A novel DC-DC convertor using LTCC technology for magnetic integration application

Z Q Xu[1,2], Y Shi[2], H P Guo[2], B C Yang[2]

[1] Research Institutes, University of Electronic Science and Technology of China, Chengdu, China

[2] State key Laboratory of Electronic Thin Films and Integrated Devices, University of Electronic Science and Technology of China, Chengdu, China

E-mail: nanterxu@uestc.edu.cn

Abstract. A compact DC-DC convertor is proposed and fabricated in multilayer ferrite substrate using LTCC technology. The spiral conductors are buried into the ferrite substrate with a multilayer 3-D structure to reduce the volume of convertor. The passive integration of magnetic components and surface circuitry are achieved in a whole substrate and the size of module can be reduced markedly. The whole height of module is only 3mm, 1/3 of the height of conventional modules. Testing results indicate that the performance of the module is excellent in Point-of-Load (POL) field. The step-down DC-DC converter converts input voltage of 5V to output voltage of 3.3V. It is confirmed that the maximum conversion efficiency of 93.2% is sufficient for actual DC-DC converter application. Such a compact DC-DC convertor provides a compact, low cost and high reliability approach for power supply and magnetic integration application.

1. Induction

The latest communication systems demand ever-greater functionality, higher performance, and lower cost in smaller and lighter formats. The power supply modules, especially DC-DC convertors, which cover significant space in the wireless communication terminals and systems [1-2]. To reduce the size and weight of DC-DC convertor is the key issue to achieve system miniaturization. Magnetics is the major bottleneck for achieving compact high power density DC-DC convertor [3-4]. Therefore, how to design a highly integrated DC-DC convertor has become a widespread interest around the world. In order to reduce the converter footprint and fully use the available space, 3D integration concepts are widely used in both areas, which design the magnetic component as a low-profile substrate [5].

There are some available integration technologies for multilayer integration applications, such as thin film technology, high-temperature co-fired ceramic (HTCC) technology, and low-temperature co-fired ceramic (LTCC) technology [6-7]. Particularly, the advantages of LTCC have motivated substantive research. In recent years, LTCC technologies are widely used for their superior advantages over other substrate technologies, such as three-dimensional (3-D) integration capabilities, small value of dielectric loss tangent and low cost for high-volume production [8]. Although LTCC technology

[1] To whom any correspondence should be addressed.

has been widely adopted in RF and microwave electronics, using LTCC technology for power electronic applications is currently a relatively new area [9].

In this paper, a novel DC-DC convertor is proposed and fabricated in LTCC technology. In order to reduce the volume of the convertor, the spiral conductors are buried into the ferrite substrate. The passive integration of magnetic components and surface circuitry are achieved in a whole substrate and the size of module can be reduced markedly. Testing results indicate that the performance of the module is excellent in Point-of-Load field. The whole height of module is only 3mm which provides a compact, low cost and high reliability approach for power electronics application.

2. Structure design

Figure 1 shows a simplified functional diagram for a step-down DC-DC converter circuit. For this configuration, we would like to fabricate 2.6uH inductor, as circled in Fig. 1. In order to reduce the volume of the convertor, the spiral inductor was adopted as the basic design in the substrate to optimize the inductance as well as conductor resistance. Because it is difficult to design a higher density of turn number in toroidal inductor. On the other hand, it is possible to design a multilayer spiral inductor. Therefore, the spiral conductors are buried into the ferrite substrate to achieve a multilayer 3-D structure. Use of Ag conductor makes it possible to achieve low resistance of the inductor pattern. The thickness of conductor is 10um. To obtain an inductance of 2.6uH, the inductor are be computed as

$$L = \frac{\varphi}{I} = \frac{\int_S B_1 dS_1}{I} + \frac{\int_S B_2 dS_2}{I} = \frac{\mu_0 \mu_r \int H_1 dS_1}{I} + \frac{\mu_0 \int H_2 dS_2}{I} \tag{1}$$

where B_1 is the magnetic flux density, and H_1 is the magnetic field strength of the magnetic core, B_2 is the magnetic flux density, and H_2 is the magnetic field strength of the air [10].

In addition, the parameters of inductor can be optimized through Electromagnetic calculation and simulation based on the computing formula of inductance. Figure 2 shows the simulation model of the multilayer inductor, which W is the line width of the conductor, D is the thickness of the ferrite substrate, and S (the sum of A, B and C) is the length of the conductor. After calculation and simulation, the line width of 0.75 mm, the thickness of 1.4 mm, and the length of 11.75mm are obtained. The LTCC inductor has no saturation and exhibits high inductance value. Then, switching loss of converters could be largely reduced for very low current ripple. Moreover, an inductor fabricated using LTCC technology is used as a substrate where the traces are printed and devices mounted on top of the substrate. In this way, the passive integration of magnetic components and surface circuitry are achieved in a whole substrate and the size of module can be reduced remarkably.

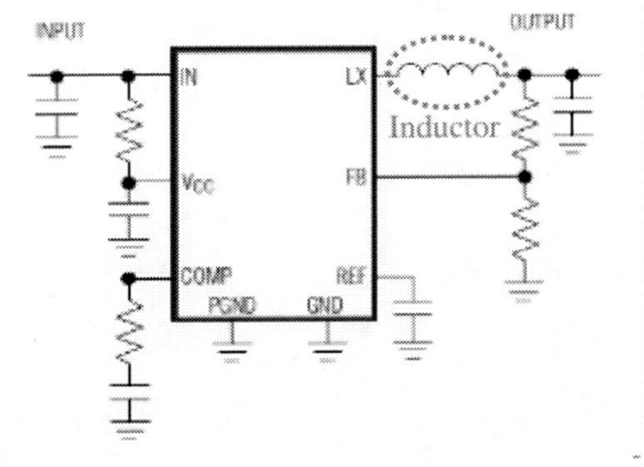

Figure 1. The functional diagram of the convertor

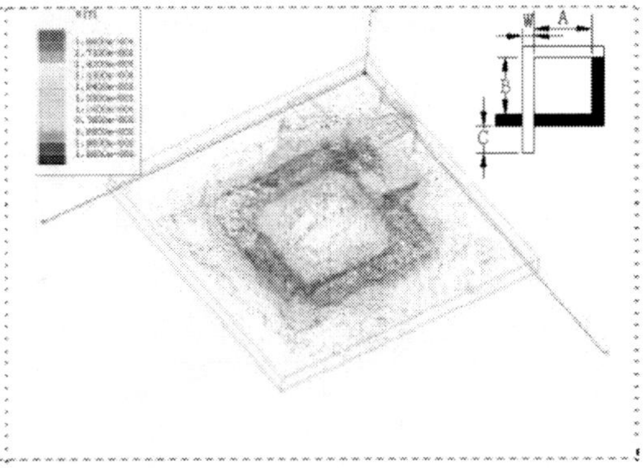

Figure 2. The simulation model of the multilayer inductor

3. Experimental process

Based on the proposed schema and 3-D multilayer structure, a DC/DC convertor was fabricated using the LTCC process, as shown in Fig. 3. The initial permeability and Q factor of the LTCC ferrite green tape (casting with PPT-LSF400 powders) are 450 and 90.

First, the ferrite green tap is cut and via holes are punched. Vias for conduction between layers and wiring patterns are screen printed on the green sheet using conductive paste. Many layers of these printed green sheets are arranged in layers, then heat and pressure is applied to laminate them (the organic resin in the green sheets acts as glue for bonding the layers during lamination). By firing the conductor metal and ceramic together while driving off organic binder in them, the DC/DC convertor in a multilayer ceramic substrate can be obtained. The measured metal's thickness of the conductor is about 10 micrometer.

Based upon the above experimental process, the sintering profile, as shown in Fig. 4, were modified in order to eliminate inner stress in the multilayer substrate. The profile has a lower temperature rising rate about 3 ℃ per minute in the range of room temperature to 280 ℃, 280 ℃ to 450 ℃, and reaches the peak temperature of 900 ℃, lasting for 0.5 h. With a properly controlled process, we can achieve excellent performance to meet desired specifications.

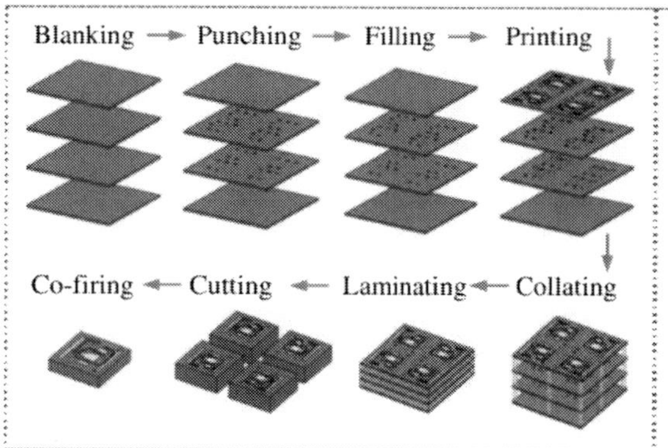

Figure 3. The fabricated process of the LTCC DC/DC convertor

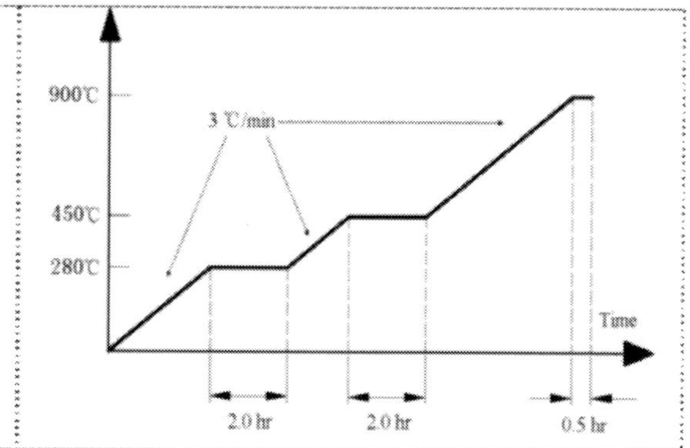

Figure 4. The sintering profile during the co-firing process

4. Experimental results

After the analysis and optimization of experimental process, the designed multilayered buck DC/DC convertor is fabricated and measured. Measurement results indicate that the performance of the module is excellent in POL field. The step-down DC-DC converter converts input voltage of 5V to output voltage (V_o) of 3.3V. Maximum output ripple (V_{pp}) of 24mV, output power (P_o) of 5W and output regulation of less than 1% are achieved to meet the design specifications. Conversion efficiency of the DC-DC converter was calculated by the ratio of input power to output power.

Table 1. The testing result of the DC-DC converter

Testing Points	R_1	R_2	R_3	R_4	R_5	R_6	R_7	R_8	R_9
P_o (W)	1.4	1.6	1.9	2.3	2.9	3.8	4.5	4.8	5.0
V_{pp} (mV)	14.1	14.2	15.1	15.8	17.0	19.2	22.8	23.7	24.0
Efficiency (%)	92.3	93.2	92.7	91.6	91.0	92.8	91.8	90.2	90.2

Table 1 shows the measurement results of the DC-DC converter. It is confirmed that the maximum conversion efficiency of 93.2% is sufficient for actual DC-DC converter application. Figure 5 shows the samples of the convertor. The convertor with integrated inductor substrate has dimensions

10mm×10mm×3.0mm. The whole height of module is only 3.0mm, 1/3 of the height of traditional modules. It is feasible to achieve miniaturization power supply modules by magnetic integration utilize LTCC technology.

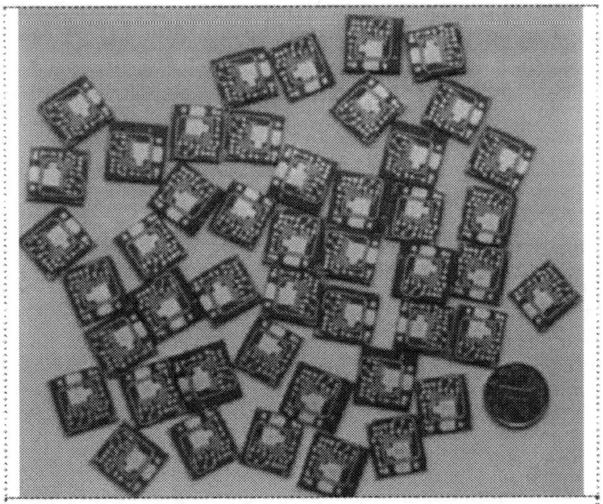

Figure 5. The samples of the convertors

5. Conclusions

This paper proposes a novel DC-DC convertor with the spiral conductor is buried into the ferrite substrate. The convertor is fabricated in multilayer ferrite substrate using LTCC technology. Testing results indicate that the performance of the module is excellent. The step-down DC-DC converter converts input voltage of 5V to output voltage of 3.3V. It is confirmed that the maximum conversion efficiency of 93.2% is sufficient for actual DC-DC converter application. Furthermore, the height of the DC/DC convertor with integrated inductor substrate is only 3mm, 1/3 of the height of traditional modules. The DC-DC convertor provides a low cost and high reliability approach for magnetic integration application

6. Acknowledgment

The authors wish to thank the Fundamental Research Funds for the Central Universities (Grant No. ZYGX2009J091), Sichuan Provincial Fundamental Research of China (Grant No. 2008JY0057) for financial support.

References

[1] Chen R T and Chen Y Y 2005 *IEE Proc. Electr. Power Appl.* 152 977.
[2] Zhou X, Donati M, Amoroso L, and Lee F C 2000 *IEEE Trans. Power Electron.* 15 826.
[3] Lim M H F and Wyk J D 2008 *IEEE Tran. Power Electro.* 23 1556.
[4] Prieto M J, Pernia A M, Lopera J M, Martin J A, and Nuno F 2002 *IEEE Trans. Ind. Applica.* 38 543.
[5] Kuo J T, Yeh T H, and Yeh C C, 2005 *IEEE Trans. Microw. Theory Tech.* 53 1331.
[6] Tsai L C and Hsue C W, 2004 *IEEE Trans. Microw. Theory Tech.* 52 1111.
[7] Lim M H F, Liang Z, and vanWyk Z D 2007 *IEEE Trans. Compon. Packag. Technol.* 30 170.
[8] Roshen W A 1990 *IEEE Trans. Magn.* 26 270.
[9] Lim M H F, vanWyk Z D, and Liang Z 2009 *IEEE Trans. Compon. Packag. Technol.* 32 3.
[10] Mulligan M D, Broach B, and Lee T H 2005 *Power Electron. Lett* 30 24.

Journal of Physics: Conference Series **263** (2011) 012011 doi:10.1088/1742-6596/263/1/012011

Design and Simulation of Interface ASIC for Micro-fluxgate Sensor

Chao Zuo Shi Chen[1] Shaobin Wu Jun Ouyang Lei Yu Xiaofei Yang
Department of Electronic Science and Technology, Huazhong University of Science and Technology, Wuhan, China

E-mail: s_chen@mail.hust.edu.cn

Abstract. Application Specific Integrated Circuits (ASIC) for micro-fluxgate sensor, based on the Complementary Metal Oxide Semiconductor (CMOS) fabrication, was designed by a 0.5μm n-well 2p-2m process, which has the advantage of small dimension and low-power consumption. In consideration of the structure and specification of the micro-fluxgate sensor, the excitation and sensing circuits were designed based on second-harmonic operation principle. The functions and parameters of circuits were verified by simulation, and the area of chip was 2.1mm×2.8mm.

1. Introduction

Fluxgate sensors measure dc or low-frequency ac magnetic fields. They are vector devices and sensitive to the field direction in the range of up to 1mT with achievable resolution down to 10 pT [1]. The first fluxgate sensors appeared in early 1930s and these sensors are still being used in many applications. However , Recent development of magnetoresistors, especially anisotropic magnetoresistance sensors (AMR), limit the market for classical fluxgate to applications requiring high precision and small size.

Small-size fluxgates are needed and used in many applications, such as magnetic ink reading or sensor arrays. But it is complicated to miniaturize the fluxgate, because the fabrication of micro-fluxgate is complex. With improvement of integration technology, the preparation of fluxgate probe and its driving circuit becomes compatible. Recently, several portable micro-fluxgate sensors with Micro-Electro-Mechanical Systems (MEMS) fabrication have been developed. The MEMS micro-fluxgate sensors keep progressing in miniaturization, integrated, systematization and intelligent with semiconductor materials technology [2]. In order to solve the traditional drive circuits' shortcomings such as complex wiring, electromagnetic crosstalk, high power consumption and large size, a chip of application specific integrated circuits (ASIC) was designed in this paper.

2. Structure of micro-fluxgate sensor

As shown in Fig.1 [1]. T The excitation current I_{ext} through the excitation coil produces field that periodically saturates (in both directions) the soft magnetic material of the sensor core. In saturation the core permeability drops down and the DC flux associated with the measured dc magnetic field B_0 is decreased. When the measured field is present, the voltage V_{ind} is induced into the sensing coil at the second and higher even harmonics of the excitation frequency. This voltage, proportional to the measured field, is the sensor output.

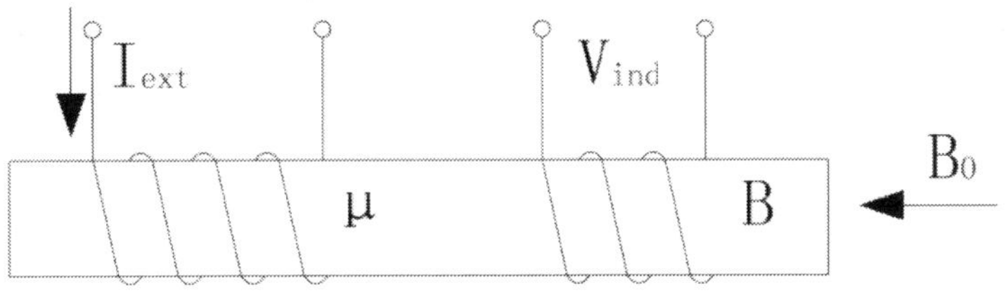

Figure 1. Fluxgate principle.

The system circuits were based on second harmonic operation principle [3], as shown in Fig.2. The square-wave oscillator and divider generated excitation current. The amplifier enhanced the sensing signal and the filter picked up the second harmonic components. After the phase-sensitive detection and low-pass filter, the output signal was directly proportional to the external magnetic field.

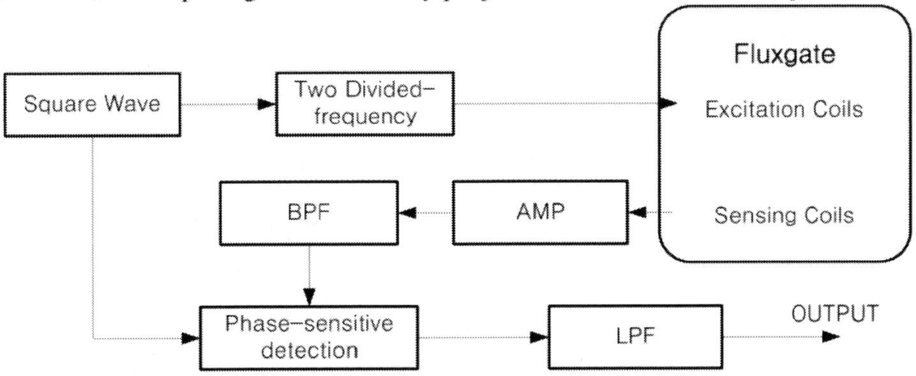

Figure 2. Block diagram of fluxgate interface circuit.

The micro-fluxgate probe was designed as a runway model's structure and consist a runway model's magnetic core based on cobalt-based amorphous alloy and two coils sputtered by copper [4], as shown in Fig.3. The excitation coils were designed as 32 turns and the sensing coils had 14 turns. The size of whole micro-fluxgate probe was 5.7mm×7.1mm×60 μ m.

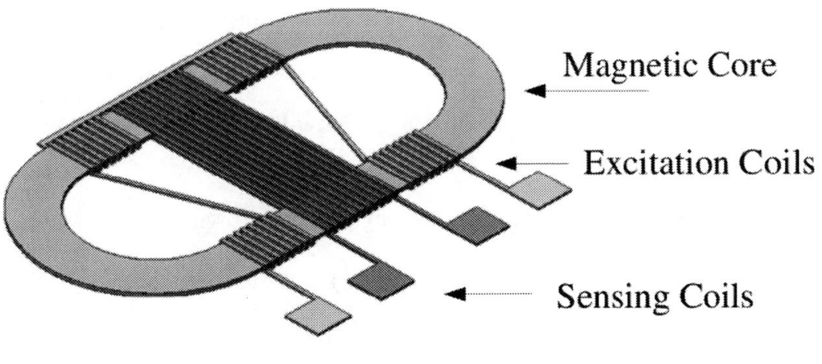

Figure 3. Structure of micro-fluxgate probe.

3. Design of interface circuit

3.1. Excitation circuit

As shown in Fig.4, G1~G3, R, Rs and C make up the RC ring oscillator to provide the square wave. Rs is the over current protection resistor and R, C are time delay components. Known by the circuit principle, the frequency can be achieved:

$$f = \frac{1}{2.2RC} \tag{1}$$

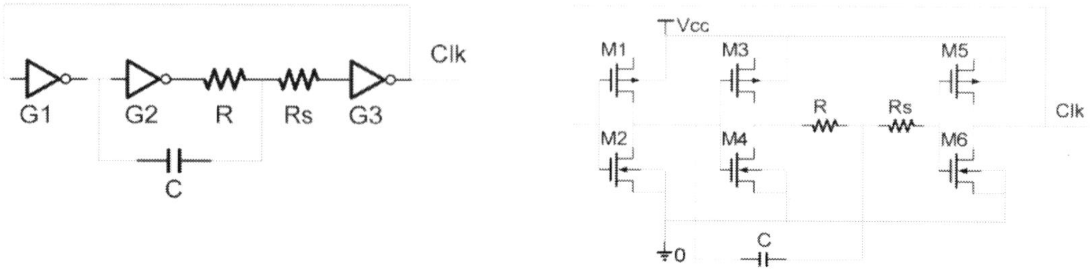

(a). Structure. **(b).** Transistor.

Figure 4. Circuit diagram of square wave.

According to design requirements, the RC parameters were adjusted to make the square wave frequency at 31.114 kHz with duty cycle of 51.612%. As shown in Fig.5, the square wave is standardized by wave-shaping circuit with inverter G4 and G5 to provide the reference signal for phase-sensitive detection. And the wave divided by D trigger is the source of excitation coils, with 15.617 kHz and duty cycle of 50.021%.

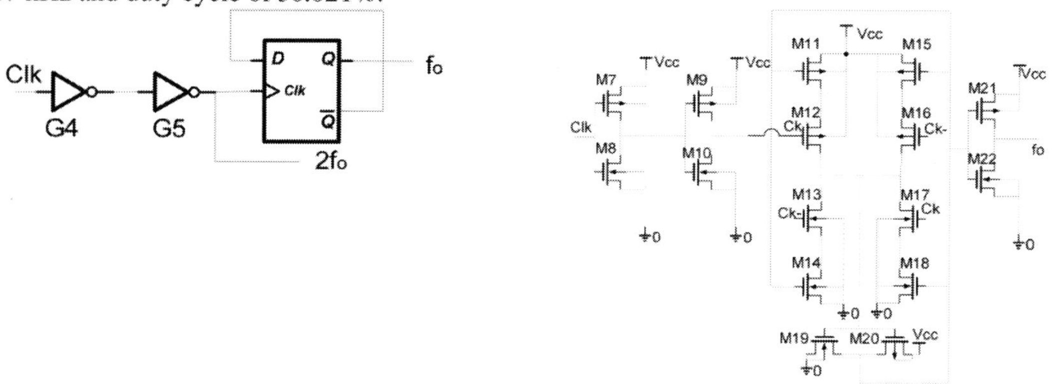

(a). Structure. **(b).** Transistor.

Figure 5. Circuit diagram of divider.

3.2. Sensing circuit

The signal of sensor is weak and a preamplifier is designed as shown in Fig.6, and the gain of differential AMP is 91.154 dB [5].

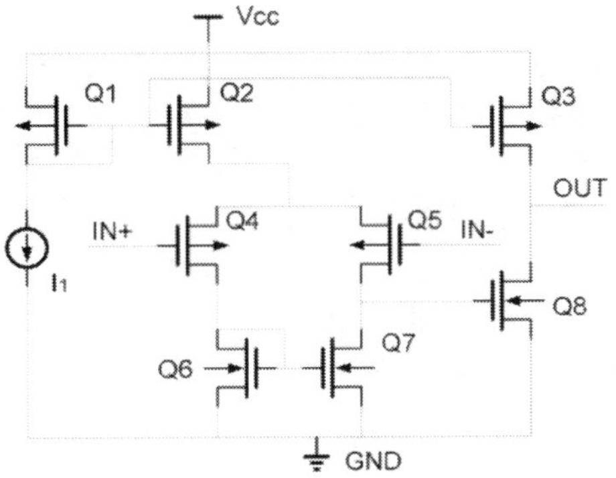

Figure 6. Circuit diagram of AMP.

A Butterworth first-order band-pass filter is designed to collect the second harmonic, as shown in Fig.7, with centre frequency at 31.212 kHz, band pass width of 4.048 kHz and quality factor to 8.276.

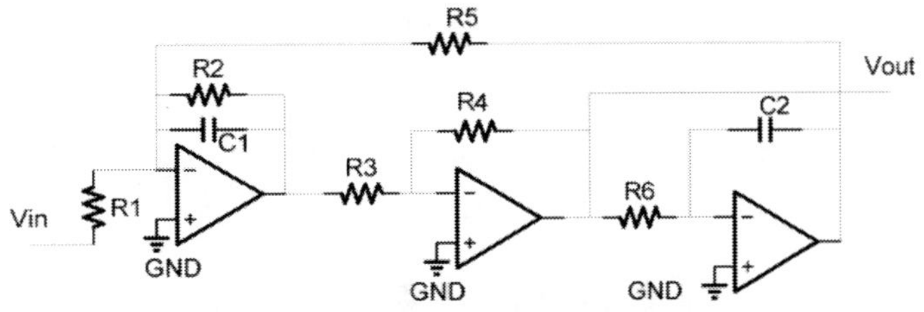

Figure 7. Circuit diagram of BPF.

As shown in Fig.8, a dual AMP and an analog switch compose the phase sensitive detection circuit. The switch phase sensitive detection circuit maintains the input waveform during the positive half-cycle and reverses the wave in the Negative half-cycle. The following second-order active low-pass filter circuit make the signal a DC output, with the cut-off frequency of 100 Hz.

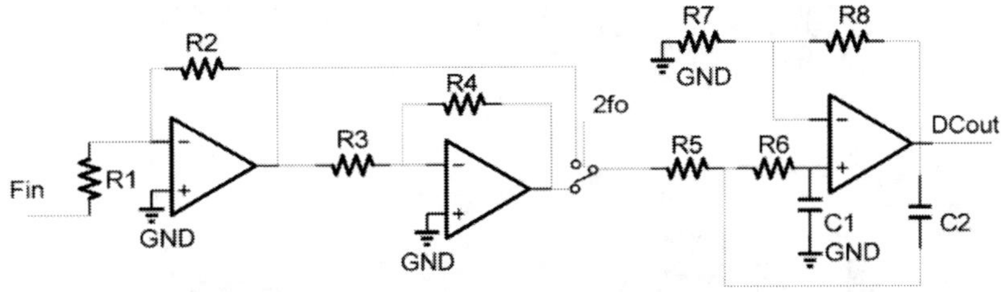

Figure 8. Circuit diagram of detection and wave-shaping.

4. Layout-design and post-simulation

With Cadence Virtuoso layout tools, the layout is designed in digital-analog mixed environment based on the model of a 0.5 μ m n-well 2p-2m process, includes 24 Pads, 254 transistors, 19 resistors and 8 capacitors. As shown in Fig.9, the whole ship takes a size of 2100×2800 μ m^2, with the digital part and analog part be separated.

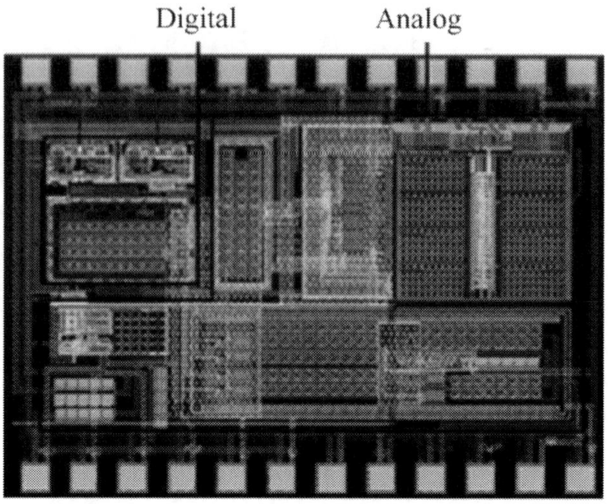

Figure 9. Layout diagram of fluxgate interface circuit.

The simulation of overall circuit has been completed after the design of micro-fluxgate interface circuit, based on standard library by HSPICE. As shown in Fig.10, the system output goes to stability in 10ms.

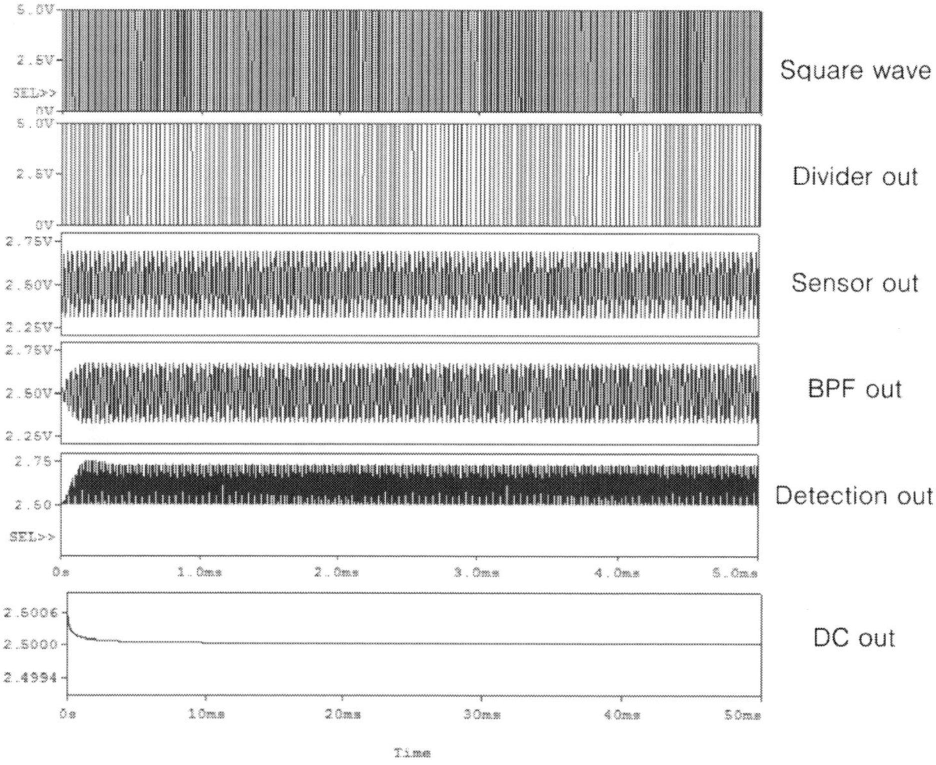

Figure 10. Simulation waveform of fluxgate interface circuit.

5. Conclusions

In this paper, a micro-fluxgate interface ASIC is designed based on t he principle of second harmonic detection of the sensing signal, and the layout and the simulation of the whole circuit have been completed. The feasibility of the circuit's function has been verified by simulation. The future work will be focused on the design of a micro-fluxgate sensor contains both the probe and circuit on one chip [6]. Because the function of MEMS-based micro-fluxgate sensor probe with runway-shaped Co-based amorphous alloy core could not be tested effectively at present, the sensor's functionalities such as sensitivity and frequency response will be discussed in a future article.

References

[1] P Ripka. Advances in fluxgate sensors. *Sensors and Actuators A*. 2003,106: 8-14.
[2] J Lenz, A S Edelstein. Magnetic Sensors and Their Applications. *Sensors and Actuators A*. 2006, 10(2):631-647
[3] Ripka P. Advances in fluxgate sensors. *Sensors and actuators A*. 2003, 106:8-14
[4] P M Wu, C H Ahn. A Fully Integrated Ring-Type Fluxgate Sensor Based on a Localized Core Saturation Method. *IEEE Transactions on Magnetics*.2007, 43(3):1040-1043
[5] Behzad Razavi. Design Considerations for Direct-Conversion Receivers .*IEEE Transactions on Circuits and Systems- II: Analog and Digital Signal Processing*. 1997, Vol.44 (No.6):22-25P
[6] A Baschirotto, E Dallago. A CMOS 2D Micro-Fluxgate Earth Magnetic Field. *2007 IEEE International Solid-State Circuits Conference*. 2007:33-38

Fabrication of MEMS-based Micro-fluxgate Sensor with Runway-shaped Co-based Amorphous Alloy Core

Shaobin Wu Shi Chen[1] Jun Ouyang Chao Zuo Lei Yu Xiaofei Yang
Department of Electronic Science and Technology, Huazhong University of Science and Technology, Wuhan, China

E-mail: s_chen@sina.com

Abstract. High-precision magnetic micro-sensor is an interdisciplinary subject of magnetic field measurement techniques and micro-electromechanical systems (MEMS) technology. A micro-fluxgate magnetic sensor based MEMS technology was designed and fabricated in this paper. This device is a micro-magnetic sensor with a symmetric construction, closed magnetic circuits and differential form. A 25μm thick Fluxgate core of runway model, made by Co-based amorphous alloy, was etched by laser and pasted on the substrate accurately. Excitation coil and sensing coil of 3D solenoid structure were prepared by RF magnetron sputtering and UV-lithography. The minimum line width of the coil is 50 μm. The experimental result shows that micro-fluxgate devices with the size of 5.7mm×7.1mm×60μm had a stable structure.

1.Introduction

Fluxgate magnetic sensor is a measurement device which are widely used in various fields. It has the features of high resolution (up to 10-11T), wide range of measureing weak magnetic fields (in the 10-8T or less), low power consumption, being able to measure the magnetic field component directly and suitable for using in high speed motion systems [1]. Micro-FluxGate is an interdisciplinary subject of magnetic field measurement techniques and micro-electromechanical MEMS technology. With the development of fluxgate sensor, micro-fluxgate sensor produced by micro-mechanical technologies such as anisotropic etching and LIGA technology, has become an inevitable development trend of fluxgate sensor

[1] *Sponsored by National Natural Science Foundation under the project (60871018)*

[2]. Because of the characteristic of micro-fluxgate, such as light weight, small size, high sensitivity, high resolution and good temperature stability and other remarkable features, it was used to be weak magnetic field measurement [3]. Micro-FluxGate sensor has important application in aerospace field, particularly in satellite technology.

Fluxgate sensor is composed with the probe and interface circuits. Probe typically includes magnetic core, excitation coil and sensing coil. Magnetic Core is generally made of soft magnetic materials with high permeability and low coercivity. Excitation coils and sensing coils are in the form of solenoid around the core, the coil material is copper [4]. For the micro-fluxgate, the core thickness and width of copper coils was tens of microns. It is difficult to electroplate the Co-based amorphous alloy on the silicon substrate, therefore the key technology is to produce solenoid coils around the thick Co-based amorphous alloy, particularly the preparation of a micro-vias column.

2.Design of Micro-fluxgate sensor

Becaused of the high permeability of the Ni-Fe permalloy, early fluxgate core commonly used it as the magnetic core. While iron-nickel can also get a high saturation magnetic flux density, and its resistance is low [5]. When the Co-based amorphous alloy soft magnetic material appears, because of its high permeability, low coercivity, low hysteresis loss, excellent wear resistance and corrosion resistance, good temperature and aging stability, it become the ideal substitute of permalloy and widely used in production of the fluxgate sensor. For those reasons, cobalt-based amorphous alloy, brand name VITVAC 6025Z, was used as the magnetic core. The relative permeability, magnetic loss, magnetostriction, electrical resistivity, squareness ratio of VITVAC 6025Z is better than Ni-Fe permalloy, and its magnetic parameters as shown below.

Table 1. magnetic parameters of permalloy and Co-based amorphous alloy

material	Relative permeability μr (1kHz) (×103)	magnetic lossPcv (kW/m3)	squareness ratio Br / Bs （%）	magnetostriction coefficient λs(×10-6)	Resistivity ρ (μΩ•cm)
80% nickel permalloy	50.0	1000	55	0	65
Co-based amorphous alloy	115.0	280	85	0	142

Making use of the runway-shaped magnetic core, with small Demagnetizing field and low vertical field induction, can improve the sensitivity of micro-fluxgate sensor. Conductor lines composed with bottom conductor, top conductor and vias [5]. The runway structure was designed as shown in figure 1.

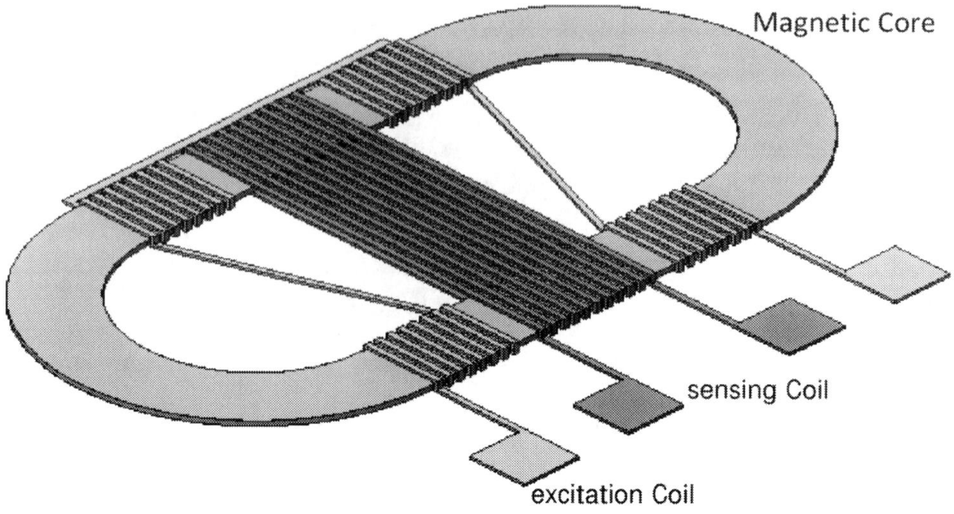

Figure 1. Structure of the Magnetic core and 3-D conductor solenoid.

The Cross section diagram was shown below. Sputtering 25 μ m thick copper on the silicon/silicon dioxide substrate, and etching it into vias column with mixed solution of hydrochloric acid and iron chloride (HCl+FeCl3). Fabricating the bottom conductor lines with lift-off method. The magnetic core was pasted on the substrate using photoresist, and it was accurately positioned between the vias column. The top conductor lines were fabricated using lift-off method again. The top conductor lines and the bottom conductor lines were connected with vias column.

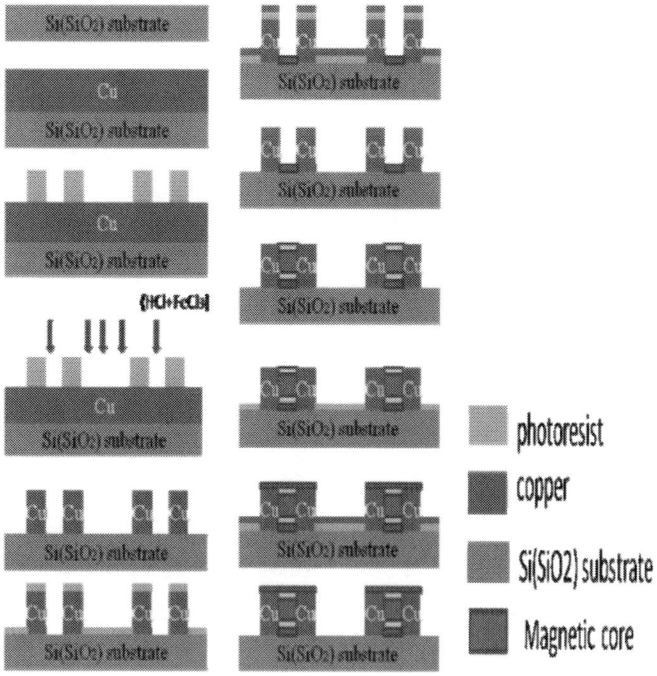

Figure 2. The fabrication processing of the Micro-fluxgate was shown in the pictures. A 25 μ m thick copper was etched to vias column in the mixed solution with HCL and FeCl3. The conductor lines made by copper was fabricatad by lift-off method. The magnetic core was pasted on the substrate using photoresist, and it was accurately positioned between the vias column.

3. Experiments & discussion

Co-based amorphous alloy was cut to runway-shaped with Laser as shown in figure 3. Magnetic hysteresis loop was tested with the VSM as shown in figure 4.

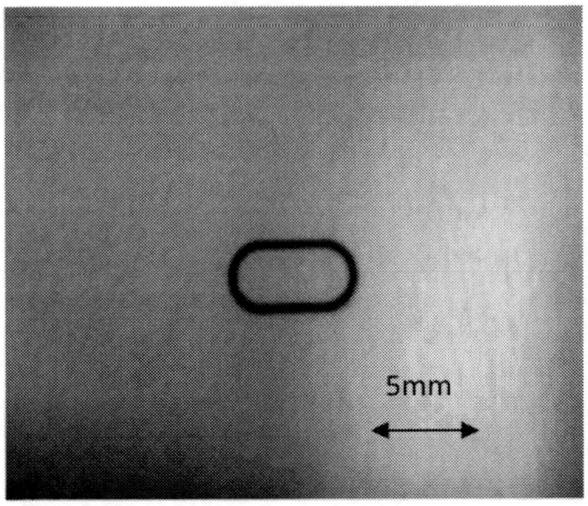

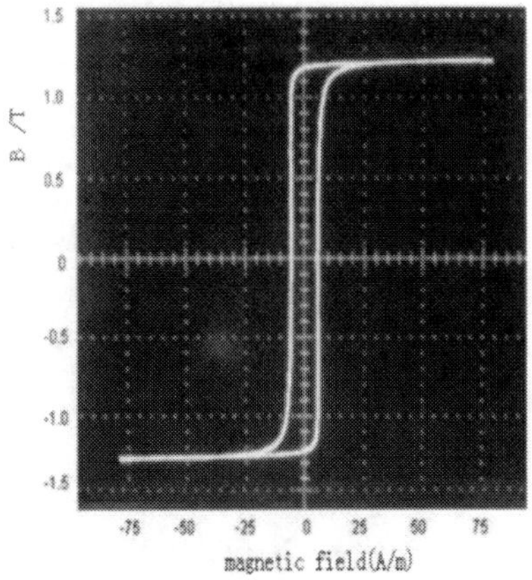

Figure 3. The runway-shaped magnetic core.

Figure 4. Magnetic hysteresis loop was tested with the VSM.

Vias of the fluxgate conductor, 25μm thick Cu, Was deposited on the silicon substrate using RF magnetron sputtering and etched to be Column as shown in figure 5. The vias column was etched by a mixed solution of hydrochloric acid and ferric chloride.

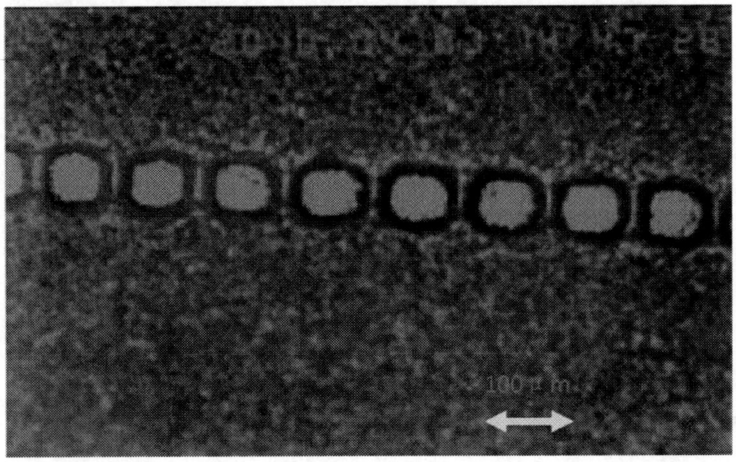

Figure 5. The vias column was etched by a mixed solution of hydrochloric acid and ferric chloride.

Bottom conductor of the fluxgate coil was prepared by the Lift-off method. Using thick photoresist (KMPE3260) and UV-lithography, the molds in the photoresist were prepared, to deposit the Ta(10nm)/Cu(1um) bottom conductors lines. Coated with KMPE3260 photoresist , exposured by ultraviolet light, the image would form a large depth to width ratio and a T-type structure as shown in figure 6.This will make it easy to complete the Lift-off.

Figure 6. The T-type structure of the KMPE3260 photoresist.

The seed layer with the metal on was then removed and a thicker photoresist layer was spun to the wafer to mask the vias. Sputtering a layer of silicon dioxide for insulation. The magnetic core and the bottom conductor were sticked together by photoresist .Also, sputtering silicon dioxide on the core and prepared top conductor, Ta20nm/Cu2μm, using lift-off Like the previous step. Then the micro-fluxgate devices was prepared as shown in figure 6. In the variable magnetic field generated by Helmholtz coils, the performance of the micro-fluxgate can be tested.

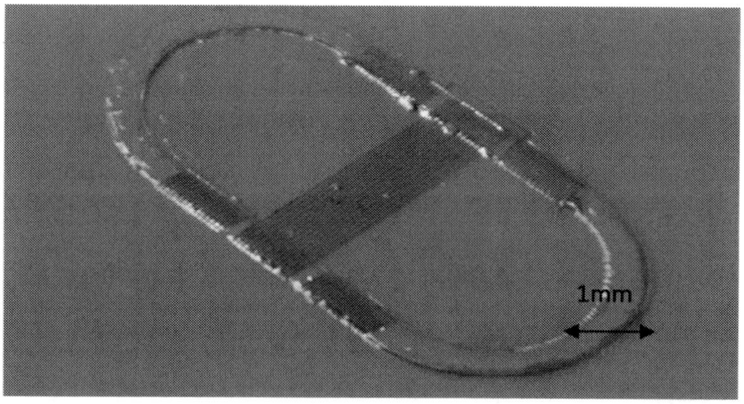

Figure 7. Surface of the device with the conductor lines and magnetic core.

4. Conclusion

Co-based Amorphous Alloy was chosen as the magnetic core first time. It was Etched by lazer into runway-shaped instead of electroplating. Compared with the permalloy, this material has a high reletive permeability, Small magnetic loss, and a perfect squareness ratio. Using UV lithography and Lift-off method to realize the miniaturization and planarization of the fluxgate. The key technology is the Preparation of three-dimensional coil using thick photoresist. The magnetic core was pasted and accurately positioned between the vias column. The experiment results show that this fabrication method was available for micro-fluxgate. The size of micro-fluxgate device was 5.7mm×7.1mm ×60μm, which can be Integrated in small devices, and it had a stable structure. Because the magnetic field measurement need specialized designed circuit, the performance of this device will be discussed in detail in a future article.

References

[1] P D Dimitropoulos, J N Avaritsiotis and E Hristoforou 2003 *Sensors and Actuators*.A **107** 238–247
[2] R Ripka 1992 *Sensors and Actuators*. A **33** 129-141
[3] R Rub, Sukirti Gupta and Chong H Ahn 2001 High Directional Sensitivities of Micro Machined Magnetic Fluxgate Sensors *Proc. IEEE Transducers'01 Conference* pp 48-151
[4] P D Dimitropoulos, J N Avaritsiotis 2003 *Sensors and Actuators*.A **106** 43–47
[5] S Kawahito, H Satoh 1996 High resolution micro-fluxgate sensing elements using closely coupled coil structures *Sensors and Actuators*. A **54** 612-617

Symmetry of Longitudinal Magneto-resistance in the Presence of Multiple Domain Walls in Ferromagnetic Thin Films with Perpendicular Magnetic Anisotropy

Gang Xiang[1], Xi Zhang
Department of Physics, Sichuan University
Chengdu, Sichuan, People's Republic of China, 610064

E-mail: gxiang@scu.edu.cn

Abstract Symmetry of the longitudinal mangetoresistance (MR) of ferromagnetic thin films with perpendicular magnetic anisotropy (PMA) in the presence of multiple domain walls (DWs) has been studied. Based on exact solutions, schematic representations of electric fields and MR have been established considering anomalous Hall effect and the effect from the eddy currents in the proximity of domain walls in the ferromagnetic samples with PMA. Symmetry analysis shows that MR with single DW in opposite sweeps is antisymmetric with respect to H=0, MR with two or 2n (n is integer, n >0) evenly distributed DWs is symmetric, and MR with the three or 2n+1 (n is integer, n >0) evenly distributed DWs is antisymmetric.

1. Introduction

Contemporary developments in spintronics [1] provide a strong motivation for understanding the interplay between electrical transport and magnetic domain walls (DWs) in both ferromagnetic metals [2,3] and in ferromagnetic semiconductors [4-6]. Although the mechanisms of various magnetoresistance (MR) effects are different, most of the intrinsic MR effects, such as the ordinary MR in nonmagnetic metals, the anistropic magnetoresistance in ferromagnetic metals [7], the giant magnetoresistance in metallic multilayers [8], and spin valve effect in ferromagnetic semiconductor multilayers [9], share the common symmetry obeying Onsager reciprocity, i.e., field dependence of longitudinal MR should be symmetric with respect to the sign of the magnetic field $\boldsymbol{H}$, i.e., $\Delta R_{xx}(\boldsymbol{H}) = \Delta R_{xx}(-\boldsymbol{H})$. However, the presence of a single domain wall, both in metallic ferromagnets and ferromagnetic semiconductors, changes this symmetry [3,5]. The physical origin of the antisymmetric term in the ferromagnets with perpendicular magnetic anisotropy (PMA) can be understood as follows [5,6]. When a DW is formed, creating a spatially inhomogeneous magnetization, the current density is perturbed by circulating currents that flow in the vicinity of the DW. This current is accompanied by an electric field essentially parallel to it, resulting in an antisymmetric Hall resistivity contribution to the longitudinal MR. However, the study on the interplay of DW and electrical transport have been focused on the effect of a single domain wall [3,5,6,10], while the effect of multiple domain walls on the longitudinal MR is rarely discussed.

In this Letter, we will discuss the symmetry of the longitudinal MR in the presence of multiple domain walls in ferromagnetic thin films with PMA. The external magnetic field is swept along easy axis [3], swept along hard axis [5] of the sample, or rotated 360^0 around one of the crystalline directions in the sample plane [5]. The electrical current flows from the left to the right, and the longitudinal MR is obtained by measuring the voltage between the electrodes α and β, as shown in Fig. 1 (a).

In our model, we treat the device as a rectangular Hall bar of width w and thickness t. Starting from the case with a single domain wall in the sample with PMA, the longitudinal resistance can be written as [6]

$$R_{xx} = R_S + R_{xy}\left[\frac{4}{\pi^2}\Sigma - \theta\left(\frac{l}{2} - |x_{DW}|\right)\right]; \tag{1}$$

$$R_S = \frac{\rho l}{wt}; \tag{2}$$

$$R_{xy} = \frac{\rho_H}{t}; \tag{3}$$

$$\Sigma = \sum_{n=odd} \frac{\exp\left(-\frac{n\pi}{w}\left|\frac{l}{2} + x_{DW}\right|\right) - \operatorname{sgn}\left(|x_{DW}| - \frac{l}{2}\right)\exp\left(-\frac{n\pi}{w}\left|\frac{l}{2} - x_{DW}\right|\right)}{n^2}; \tag{4}$$

where l is the distance between the two electrodes, x_{DW} is the domain wall location.

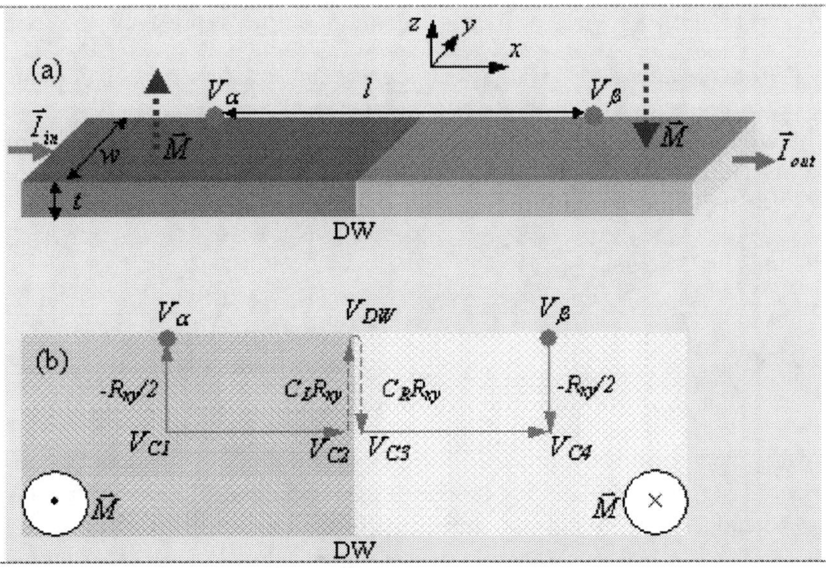

Figure 1. The schematics of a thin film with 180^0 domain wall at x =0. (a) The width and the thickness of the film are w and t. The distance between the two electrodes α and β is l. The magnetization in the left region points upward, while that in the right region downward. (b) 2-dimensional top view of the thin film sample. The direction of the red arrows represents the direction of the electric fields between different points. The dashed lines represent the electric fields around the proximity of the DW considering the effect of eddy currents. V_{C1},V_{C2}, V_{C3}, and V_{C4} are the potentials of the points along the middle line of the width of the sample. V_{C2} and V_{C3}, the potentials of the points in the proximity of the DW. V_{DW}, the potential of the DW on the upper side of the sample. The magnetization in the left region points outward, while that in the right region inward.

When the domain wall is located outside the two electrodes, but not in the proximity of the DW, we get the trivial solution, i.e., $R_{xx}= R_S$. When the single domain wall is located between the two electrodes, i.e., $|x_{DW}|<l/2$, we get the longitudinal resistance between two electrodes α and β from Eqn (1)-(4)

$$R_{\alpha\beta} = R_S + (-\frac{1}{2})R_{xy} + C_L R_{xy} + C_R R_{xy} + (-\frac{1}{2})R_{xy}; \tag{5}$$

where

$$C_L = \left(\frac{4}{\pi^2} \sum_{n=odd} \frac{\exp\left(-\frac{n\pi}{w}(\frac{l}{2}+x_{DW}) \right)}{n^2} \right) > 0;$$
$$C_R = \left(\frac{4}{\pi^2} \sum_{n=odd} \frac{\exp\left(-\frac{n\pi}{w}(\frac{l}{2}-x_{DW}) \right)}{n^2} \right) > 0; \tag{6}$$

Note that the $\sum(x)$ function expresses the perturbation of electrical current in the domain wall proximity, while the term ($-\frac{1}{2}R_{xy}$) in the equation above represents the contribution to the longitudinal MR from the anomalous Hall effect due to magnetization in each domain.

We now develop a strategy to schematically represent the longitudinal resistance terms in Equation (5) in the presence of the domain wall in the ferromagnetic samples with PMA, via connecting the exact solutions to the effective electric fields between different points in the sample.

Starting from Ohm's law, we get

$$R_{\alpha\beta} = \frac{V_\alpha - V_\beta}{I} = \frac{(V_\alpha - V_{C1}) + (V_{C1} - V_{C2}) + (V_{C2} - V_{DW}) + (V_{DW} - V_{C3}) + (V_{C3} - V_{C4}) + (V_{C4} - V_\beta)}{I} \tag{7}$$

where V_{C1}, V_{C2}, V_{C3}, and V_{C4} are the potentials of the points along the middle line of the width of the sample, as shown in Figure 1. V_{C2} and V_{C3} represent the potentials of the points in the proximity of the DW. V_{DW} is the potential of the DW on the upper side of the sample.

Note that

$$R_S = \frac{(V_{C1} - V_{C2}) + (V_{C3} - V_{C4})}{I}; \tag{8}$$

Anomalous Hall effect requires that

$$\frac{-R_{xy}}{2} = \frac{(V_\alpha - V_{C1})}{I} = \frac{(V_{C4} - V_\beta)}{I}; \tag{9}$$

By symmetry,

$$C_L R_{xy} = \frac{V_{DW} - V_{C2}}{I}; \tag{10}$$

$$C_R R_{xy} = \frac{V_{C3} - V_{DW}}{I} \tag{11}$$

Combining equations (5) – (11), we can establish the relationships between the resistance terms in the Equation (5) and the effective electric fields between different points in the sample. Since longitudinal MR is $\Delta R_{xx} = \Delta R_{ab} = (R_{ab} - R_s)/R_s$, and the value of R_S is constant as long as the distance l between points α and β is chosen, we'll focus on the other 4 terms in equation (5) only. As shown in Fig.1 (b), the arrows are used to represent the directions of effective electric fields between different points, while the magnitude of the fields is marked by the coefficients besides. The magnitude of the electric current I is omitted in the coefficients for simplicity.

Specifically, for the situation that the DW is located at the middle of the two electrodes α and β, the diagram is shown in Fig 1(b). By symmetry, we get

$$(C_L)_a = (C_R)_a \tag{12}$$

This is exactly what equation (4) shows when the domain wall is located at the middle, i.e., x_{DW} =0.

2. Single Domain Wall

We have established the schematic representation of the electric fields and longitudinal MR in different parts of the sample considering the anomalous Hall effect and the effect from the eddy currents in the proximity of the DWs. Now we can draw similar diagrams when the DW is located at other places.

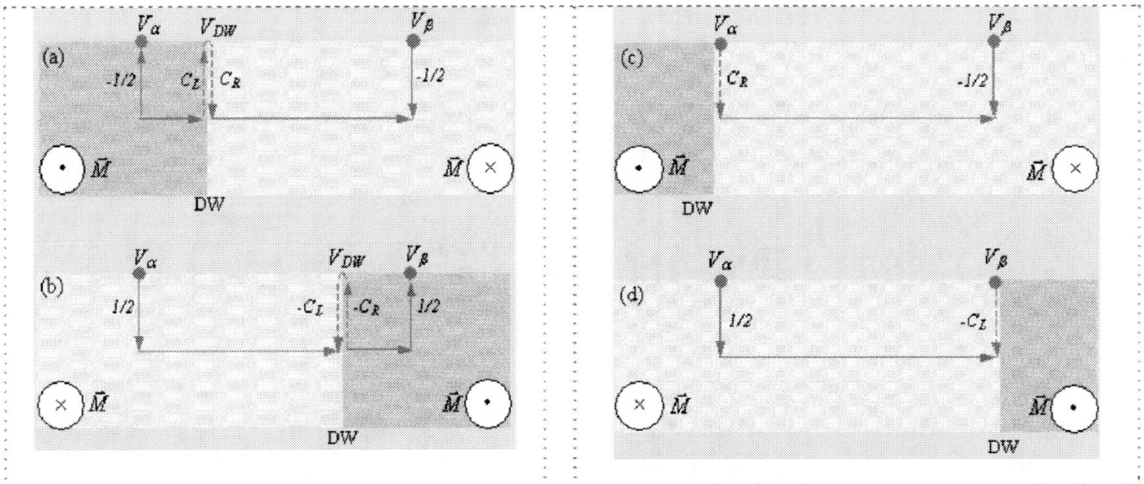

Figure 2. The schematic of a thin film with PMA in the presence of a single domain wall. (a) The DW is located between the two electrodes, closer to the left electrodes α. The magnetization in the left region is outward, while that in the right region inward. (b) The counterpart of (a). The DW is located between the two electrodes, closer to the right electrodes β. The magnetization in the left region is inward, while that in the right region outward. (c) The DW is located outside the electrodes, but in the proximity of the left electrodes α. (d) The counterpart of (c). The domain wall is located outside the electrodes, but in the proximity of the right electrodes β.

Fig 2 (a) shows the schematics when the DW is located between the two electrodes, closer to, but not in the proximity of the left electrode α. The DW configuration in Fig. 2 (b) is the counterpart of that in Fig. 2 (a): the distance between the DW and the electrode β is the same as that between the DW and the electrode α in Fig. 2(a), while the magnetization in the left and right region in Fig. (b) is opposite to that in Fig. 2(a), respectively. The configuration in Fig. 1(b) is obtained in an external ***H*** field sweep opposite to that in Fig. (a). Note that the DW configuration can always find its counterpart when the external magnetic field sweep is reversed. The value of anomalous Hall effect coefficient R_{xy} is omitted in the coefficients for simplicity.

By symmetry, the coefficients are related to each other as follows,

$$(C_L)_a = (C_R)_b \tag{13}$$

$$(C_R)_b = (C_L)_a \tag{14}$$

The coefficients C_L and C_R, express the perturbation of electrical currents in the domain wall proximity.

Since we have already connect the exact solutions to the effective electric fields between different points in the sample, now we can just go along the electric lines from V_α to V_{DW} to V_β to calculate the resistance of the system. Then for the situation in Fig. 2(a) and (b)

$$\begin{aligned}(R_{\alpha\beta} - R_S)_a &= -\frac{1}{2}R_{xy} + (C_L)_a R_{xy} + (C_R)_a R_{xy} - \frac{1}{2}R_{xy} \\ &= R_{xy}\left(-1 + (C_L)_a + (C_R)_a\right)\end{aligned} \tag{15}$$

$$\begin{aligned}(R_{\alpha\beta} - R_S)_b &= \frac{1}{2}R_{xy} - (C_R)_b R_{xy} - (C_L)_b R_{xy} + \frac{1}{2}R_{xy} \\ &= R_{xy}\left(1 - (C_L)_a - (C_R)_a\right)\end{aligned} \tag{16}$$

Thus,

$$(R_{\alpha\beta} - R_S)_a = -(R_{\alpha\beta} - R_S)_b \tag{17}$$

As a result, the longitudinal MR in this situation is antisymmetric with respect to ***H*** = 0.

When the DW is located outside the electrodes, but in the proximity of one of the electrodes, the diagrams are shown in Fig. 2 (c) and (d). Similarly, we have

$$(C_R)_c = (C_L)_d \tag{18}$$

$$(R_{\alpha\beta} - R_S)_c = -(C_R)_c R_{xy} - \frac{1}{2}R_{xy} \tag{19}$$

$$(R_{\alpha\beta} - R_S)_d = \frac{1}{2}R_{xy} + (C_L)_d R_{xy} \tag{20}$$

Therefore,

$$(R_{\alpha\beta} - R_S)_c = -(R_{\alpha\beta} - R_S)_d \tag{21}$$

As a result, the longitudinal MR in the reverse sweeps in this situation is antisymmetric with respect to ***H*** = 0.

One can do the similar symmetry analysis when the DW is located inside the electrodes but in the proximity of one of the electrodes. The same symmetry holds. When the DW is located outside the electrodes, but far away from both of them, such that the circulating current around the DW is negligible, we get $(R_{\alpha\beta} - R_S) = 0$. This is a trivial case same as a single domain sample without any DW.

In summary, in the presence of a single domain wall in the ferromagnetic sample with PMA, the analysis of symmetry in the diagram shows that when the ***H*** sweep reverse its direction, the magnetization in each domain reverses too, and the electric potential difference between difference points changes sign. This results in an antisymmetric longitudinal MR with respect to *H*=0 when the ***H*** sweeps reverse its direction, which has been observed experimentally in both metallic ferromagnets and ferromagnetic semiconductors. The similar analysis can also clearly explain the puzzling antisymmetry in ΔR_{xx} when the external field was rotated 360^0 around one of the crystalline direction in the sample plane with the same magnitude [5], but the explanation of multiple peaks in some ΔR_{xx} will have to addressed section 4. This analysis gives a detailed validation of the schematic analysis of the symmetry in MR in the presence of DW in the ferromagnetic samples with PMA.

3. Two Domain Walls

Using the similar schematic analysis shown in the Fig.3, one can do the symmetry analysis in the situation that there are two domain walls in the ferromagnetic thin films with PMA.

When the DWs are located between α and β, the diagrams are shown in Fig 3 (a) and (b) with the assumption that the distance d between the 2DWs is much shorter than the distance l between the two electrodes. Note that the DW location can always find its symmetric counterpart when the external magnetic field is swept in an opposite direction. Similarly, we use the coefficients C_{Li} and C_{Ri} to express the perturbation of electrical currents in the ith domain wall proximity. To get the expressions of C_{Li} and C_{Ri}, one needs to solve the problem exactly in the presence of two DWs, which is very difficult when there is more than one DW in the system. However, no matter what the expressions are, symmetry analysis shows the coefficients are related to each other as follows,

$$(C_{Li})_a = (C_{Ri})_b \tag{22}$$

$$(C_{Ri})_a = (C_{Li})_b \tag{23}$$

where i = 1,2. The coefficients C_{Li} and C_{Ri} expresses the perturbation of electrical currents in the ith domain wall proximity.

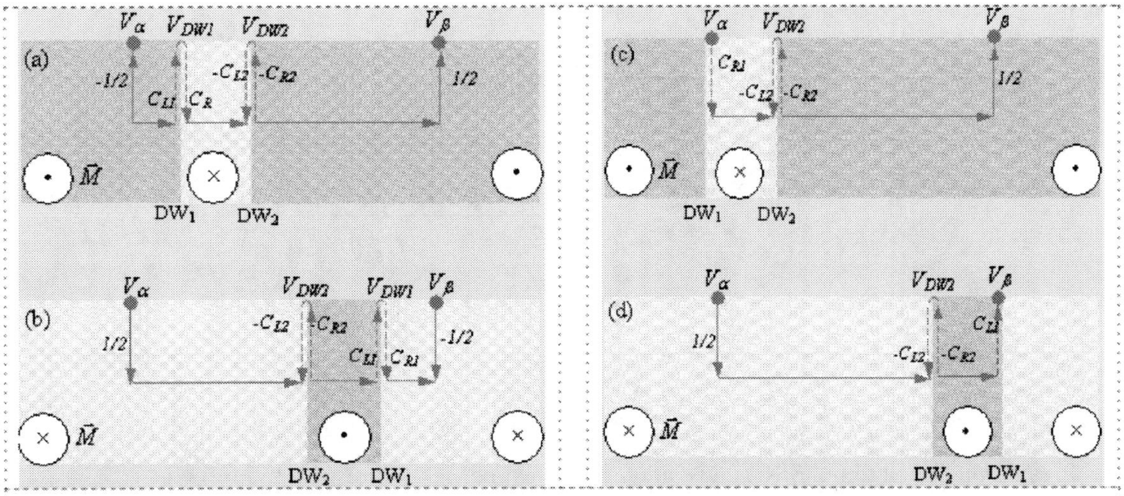

Figure 3. The schematic of a thin film with PMA in the presence of two DWs. (a) Both DWs are located between the two electrodes, closer to the left electrodes α. The magnetization in the left region is outward, while that in the right region inward. (b) The counterpart of the case in (a). Both DWs are located between the two electrodes, closer to the right electrodes β. The magnetization in the left region is inward, while that in the right region outward. (c) One DW is located outside the electrodes, but in the proximity of the left electrodes α. (d) The counterpart of the case in (c). One domain wall is located outside the electrodes, but in the proximity of the right electrodes β.

Now we can just go along the electric line from V_α to V_{DW1} to V_{DW2} to V_β to calculate the resistance of the system. Then for the situation in Fig. 2(a) and (b)

$$(R_{\alpha\beta}-R_S)_a = R_{xy}\left(-\frac{1}{2}+(C_{L1})_a+(C_{R1})_a-(C_{L2})_a-(C_{R2})_a+\frac{1}{2}\right)$$
$$= R_{xy}\left((C_{L1})_a+(C_{R1})_a-(C_{L2})_a-(C_{R2})_a\right) \tag{24}$$

$$(R_{\alpha\beta}-R_S)_b = R_{xy}\left(\frac{1}{2}-(C_{L2})_b-(C_{R2})_b+(C_{L1})_b+(C_{R1})_b-\frac{1}{2}\right)$$
$$= R_{xy}\left((C_{L1})_a+(C_{R1})_a-(C_{L2})_a-(C_{R2})_a\right) \tag{25}$$

Thus,

$$(R_{\alpha\beta}-R_S)_a = (R_{\alpha\beta}-R_S)_b \tag{26}$$

As a result, the longitudinal MR in this situation is symmetric with respect to $\boldsymbol{H}=0$.

Specifically, when the two domain walls are located equally away from the middle point of the two electrodes, the perturbation effects of the two domain walls cancel each other based on symmetry analysis, and we get

$$R_{\alpha\beta} = R_S \tag{27}$$

When one of the DWs is located outside the electrodes, and is in the proximity one of electrodes, the diagrams are shown in Fig. 3 (c) and (d). Similarly, we have

$$(C_{R1})_c = (C_{L1})_d \tag{28}$$

$$(C_{L2})_c = (C_{R2})_d \tag{29}$$

$$(C_{R2})_c = (C_{L2})_d \tag{30}$$

Then

$$(R_{\alpha\beta}-R_S)_c = R_{xy}\left((C_{R1})_c-(C_{L2})_c-(C_{R2})_c+\frac{1}{2}\right)$$
$$= R_{xy}\left(\frac{1}{2}+(C_{R1})_c-(C_{L2})_c-(C_{R2})_c\right) \tag{31}$$

$$(R_{\alpha\beta}-R_S)_d = R_{xy}\left(\frac{1}{2}-(C_{L2})_d-(C_{R2})_d+(C_{L1})_d\right)$$
$$= R_{xy}\left(\frac{1}{2}+(C_{R1})_c-(C_{L2})_c-(C_{R2})_c\right) \tag{32}$$

Therefore,

$$(R_{\alpha\beta}-R_S)_c = (R_{\alpha\beta}-R_S)_d \tag{33}$$

As a result, the longitudinal MR in this situation is symmetric with respect to $\boldsymbol{H}$=0.

One can do similar analysis for the other cases, such as or both DWs are located inside the electrodes but one of them is in the proximity of one of the electrodes, one DW is located outside the electrodes but not in the proximity of either electrodes, both DWs both are located outside the electrodes and so on, same symmetry for the MR still holds.

In summary, in the presence of two domain walls in the ferromagnetic sample with PMA, the analysis of symmetry based on the diagram shows the longitudinal MR is an even function of $\boldsymbol{H}$ and $\boldsymbol{M}$ during the opposite $\boldsymbol{H}$ sweeps. Specifically, in the special configuration in that the two domain walls are located between the two electrodes α and β and the $x=0$ plane is at the middle of the two electrodes, the longitudinal resistance $R_{ab} = R_S$. More generally, if there are 2n (n is integer, n >0) domain walls and all the domain walls are separated with the same distance, the longitudinal MR is still symmetric with respect to H=0 during the opposite ***H*** sweeps. The symmetry of MR in the presence of 2 DWs and MR without DWs share the same symmetry, but MR in the presence of 2 DWs should show extra peaks in the feature due to the resistance change when each DW passes the electrodes.

4. Three Domain Walls

In the situation that there are three domain walls in the ferromagnetic thin films with PMA, one can do the symmetry analysis using the similar schematic analysis shown in the Fig.4.

To grab the physical nature and simplify the discussion, let's start with the simplest case that the three domain walls are equally distributed. We also assume that the distance *d* between two nearest 2DWs is much shorter than the distance *l* between the two electrodes. When the DWs are located between α and β, the diagrams are shown in Fig 3 (a) and (b). Note that the DW location can always find its symmetric counterpart when the external magnetic field is swept. Similarly, we use the coefficients C_{Li} and C_{Ri} to express the perturbation of electrical currents in the *i*th domain wall proximity. To get the expressions of C_{Li} and C_{Ri}, one needs to solve the problem quantitatively in the presence of three DWs. By symmetry, the coefficients are related to each other as follows,

$$(C_{Li})_a = (C_{Ri})_b \tag{34}$$

$$(C_{Ri})_a = (C_{Li})_b \tag{35}$$

where i = 1,2,3.

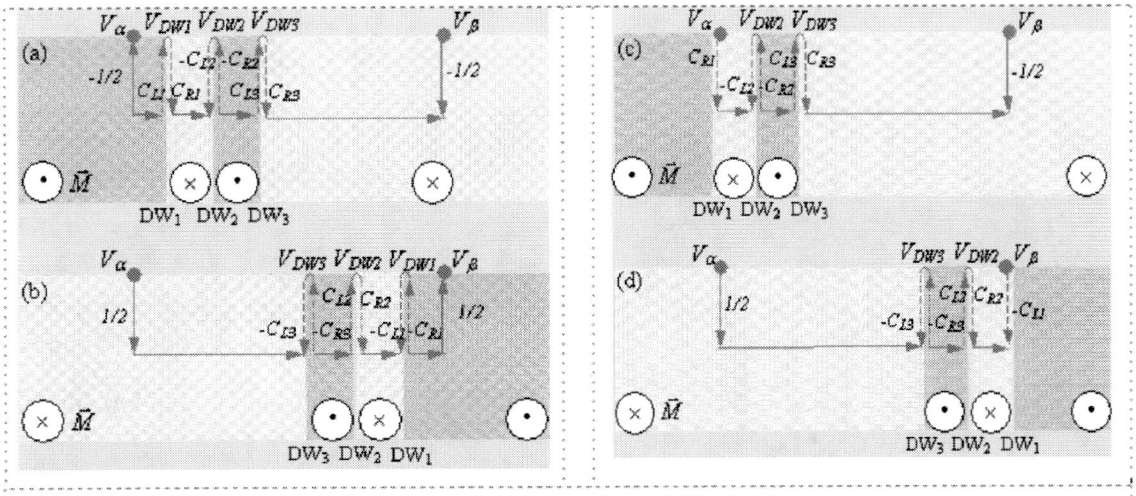

Figure 4. The schematic of a thin film with PMA in the presence of three DWs. (a) all the DWs are located between the two electrodes, closer to the left electrodes α. The magnetization in the left region is outward, while that in the right region inward. (b) The counterpart of the case in (a). all the DWs are located between the two electrodes, closer to the right electrodes β. The magnetization in the left region is inward, while that in the right region outward. (c) One DW is located outside the electrodes, but in the proximity of the left electrodes α. (d) The counterpart of the case in (c). One domain wall is located outside the electrodes, but in the proximity of the right electrodes β.

Then for the situation in Fig. 3(a) and (b)

$$(R_{\alpha\beta}-R_S)_a = R_{xy}\left(-\frac{1}{2}+(C_{L1})_a+(C_{R1})_a-(C_{L2})_a-(C_{R2})_a+(C_{L3})_a+(C_{R3})_a-\frac{1}{2}\right)$$
$$= R_{xy}\left(-1+(C_{L1})_a+(C_{R1})_a-(C_{L2})_a-(C_{R2})_a++(C_{L3})_a+(C_{R3})_a\right) \quad (36)$$

$$(R_{\alpha\beta}-R_S)_b = R_{xy}\left(\frac{1}{2}-(C_{L3})_b-(C_{R3})_b+(C_{L2})_b+(C_{R2})_b-(C_{L1})_b-(C_{R1})_b+\frac{1}{2}\right)$$
$$= R_{xy}\left(1-(C_{L1})_a-(C_{R1})_a+(C_{L2})_a+(C_{R2})_a-(C_{L3})_b-(C_{R3})_b\right) \quad (37)$$

Thus,

$$(R_{\alpha\beta}-R_S)_a = -(R_{\alpha\beta}-R_S)_b \quad (38)$$

As a result, the longitudinal MR in this situation is antisymmetric with respect to $\boldsymbol{H}$=0.

When one of the DWs is located outside the electrodes, and is in the proximity of one of the electrodes, the diagrams are shown in Fig. 3 (c) and (d). Similarly, we have

$$(C_{R1})_c = (C_{L1})_d \quad (39)$$

$$(C_{Li})_c = (C_{Ri})_d \quad (40)$$

$$(C_{Ri})_c = (C_{Li})_d \quad (41)$$

where i = 2, 3. Then by going along the electric line from V_α to V_β, we get

$$(R_{\alpha\beta}-R_S)_c = R_{xy}\left((C_{R1})_c-(C_{L2})_c-(C_{R2})_c+(C_{L3})_c+(C_{R3})_c-\frac{1}{2}\right) \quad (42)$$

$$(R_{\alpha\beta}-R_S)_d = R_{xy}\left(\frac{1}{2}-(C_{L3})_d-(C_{R3})_d+(C_{L2})_d+(C_{R3})_d-(C_{L1})_d\right)$$
$$= -R_{xy}\left((C_{R1})_c-(C_{L2})_c-(C_{R2})_c+(C_{L3})_c+(C_{R3})_c-\frac{1}{2}\right) \quad (43)$$

Therefore,

$$(R_{\alpha\beta}-R_S)_c = -(R_{\alpha\beta}-R_S)_d \quad (44)$$

As a result, the longitudinal MR in this situation is symmetric with respect to $\boldsymbol{H}$ =0.

One can do similar analysis for the other cases, such as all the DWs are located inside the electrodes but one of the DWs is in the proximity of one of the electrodes, two of the DWs located outside the electrodes, three DWs located outside the electrodes and so on, same symmetry for the MR still holds.

In summary, when there are three domain walls in the ferromagnetic thin films with perpendicular anisotropy, detailed symmetry analysis based on the diagram shows that that the longitudinal MR is antisymmetric with respect to H=0 during the opposite $\boldsymbol{H}$ sweeps. To get the exact expressions of longitudinal MR one needs the exact solution of the problem in the three-domain-wall system. More generally, if there are 2n+1 (n is integer, n >0) domain walls and all the domain walls are separated evenly, the longitudinal MR is still antisymmetric with respect to H=0 during the opposite $\boldsymbol{H}$ sweeps.

Although the symmetry of MR in the presence of single DW and MR in the presence of odd numbers of DWs share the same symmetry, there is one major difference: the MR in the presence of multiple DWs should have multiple peaks in the MR feature due to the abrupt changes of MR when each DW passes the electrodes. The results here could be used to explain the multiple peaks in the antisymmetry of ΔR_{xx} spikes when the external field was rotated 360^0 around one of the crystalline direction in the sample plane with the same magnitude [5].

5. Conclusion

The effect of multiple domain walls on the longitudinal MR of the ferromagnetic samples with PMA has been studied. By connecting the electric fields in the sample to the anomalous Hall effect and the effect from the eddy currents in the proximity of the DWs, the schematic representation of the electric fields and longitudinal MR in different parts of the sample has been established. Based on the symmetry analysis of the schematics, the symmetry of the ferromagnetic thin films with PMA with multiple domain walls are discussed. The results could be useful to fundamental studies of MR in ferromagnetic samples, and be potentially used in the magnetic devices that incorporate domain wall generation, manipulation and detection.

Acknowledgement. This work was financially supported by National Natural Science Foundation of China (Grant 11004141 and Grant 11004142).

References

[1] S. A. Wolf, D. D. Awschalom, R. A. Buhrman, J. M. Daughton, S.von Molnar, M. L. Roukes, A. Y. Chtchelkanova, and D. M. Treger, *Science* **294**, 1488 (2001)
[2] D. A. Allwood *et al.*, *Science* **296**, 2003 (2002)
[3] X. M. Cheng, S. Urazhdin, O. Tchernyshyov, C. L. Chien, V. I. Nikitenko, A. J. Shapiro, and R. D. Shull, *Phys. Rev. Lett.* **94**, 017203 (2005)
[4] H. X. Tang, S. Masmanidis, R. K. Kawakami, D. D. Awschalom, and M. L. Roukes, *Nature* **431**, 52 (2004).
[5] G. Xiang, A. W. Holleitner, B. L. Sheu, F. Mendoza, O. Maksimov, M. B. Stone, D. D. Awschalom, P. Schiffer and N. Samarth, *Phys Rev. B Rapid Comm.* **71**,241307 (R)(2005)
[6] G. Xiang and N. Samarth, *Phys. Rev. B* **76**, 054440 (2007)
[7] R. I. Potter, *Phys. Rev. B* **10**, 4626, (1974).
[8] M. N. Baibich *et al.*, *Phys. Rev. Lett.* **61**, 2472 (1988).
[9] G. Xiang, B. Sheu, M. Zhu, P. Schiffer and N. Samarth, *Phys. Rev. B* **76**, 035324 (2007)
[10] W. Desrat, S. Kamaral, F. Terki, S. Charar, J. Sadowski and D. K. Maude, *Semicond. Sci. Technol.* **24,** 065011 (2009)

Journal of Physics: Conference Series **263** (2011) 012014 doi:10.1088/1742-6596/263/1/012014

Ferromagnetic ordering in $EuCu_2P_2$

Dexuan Huo[1, 2, a], **Jun Lin**[1], **Guanghong Tang**[1], **Lingwei Li**[1], **Zhenghong Qian**[2], **and Toshiro Takabatake**[3]

[1] Institute of Materials Physics, Hangzhou Dianzi University, Hangzhou 310018, China
[2] Center for Spintronics and Device, Hangzhou Dianzi University, Hangzhou 310018, China
[3] Department of Quantum Matter, ADSM, Hiroshima University, Higashi-Hiroshima 739-8530, Japan
E-mail: dxhuo@hdu.edu.cn

Abstract. The crystal structural, transport and magnetic properties of $EuCu_2P_2$ were investigated. A ferromagnetic transition around 51 K was confirmed by examining the temperature dependence of magnetization and electrical resistivity. The results were discussed and compared with the isostructural compounds EuT_2P_2 (T = Fe, Co, and Ni). Depending on the type of transition metal, the compound shows ferromagnetic、antiferromagnetic、mixed valence and ferromagnetic ground state for Eu when the transition metal changes from Fe to Cu. The results demonstrate that the distance between the Eu ions and the P-P distance between the T_2P_2 layers play an important role in the magnetism of the compounds.

1. Introduction

The intermetallic compounds EuT_2P_2 (T = Fe, Co, Ni, and Cu) belong to an interesting group of materials with a $ThCr_2Si_2$-type structure, which show various physical phenomena due to the structural instability and competition between RKKY interaction and Kondo effect [1]. The mixing of 4f state with conduction-band states lead to the formation of different ground states in these compounds. For example, $CeCu_2Si_2$ shows a very large specific heat coefficient and superconducting transition [2]. In this so-called heavy fermon superconductor, the magnetic moment of Ce^{3+} disappears. For some Yb- and Eu-based compounds, valence transitions are observed because of the interconfigurational fluctuation [3, 4]. These ground states can also be tuned by physical/chemical pressure and magnetic field. The pressure-induced superconductivity in rare-earth magnetic compounds is a hot topic in the field of condensed matter physics [5, 6].

Recently, $EuFe_2As_2$ are attracting much attention due to the discovery of Fe-based superconductivity [7, 8]. $EuFe_2P_2$ is a rare Eu-containing dense Kondo lattice compound [9, 10]. Comparing to the extensive studies rare earth silicides and germanides, the research works on ternary pnictides RT_2Pn_2 (Pn = P, As, Sb) are less.

[a] To whom any correspondence should be addressed.

Here, we give a report on the crystal structural, transport and magnetic properties $EuCu_2P_2$. The Eu is confirmed to be in divalent valence state in our measured temperatures, showing a ferromagnetic ordering transition at 51 K.

2. Experimental

The samples of $EuCu_2P_2$ were synthesized by heating very slowly reactant of Ba (99.9%), Cu (99.999%), and P (99.999%) in evacuated fused silica tube. It took 2 days to heat the reactant to 500 ℃ gradually to avoid explosion of phosphor. Then, the temperature was raised to 850 ℃ and kept there for 24 h. The sample of powder with single crystals in micrometer size was obtained after cooling down to room temperature naturally. The structural properties and phase purity were examined by powder x-ray diffraction (XRD) and electron-probe microanalysis (EPMA). The polycrystalline samples were obtained after cold-pressing and annealing of the powder. Electrical resistivity was measured by a standard four-probe method from 8 to 300 K with home-made cryostat. The magnetic measurement was performed on a vibrating sample magnetometer (VSM) which is an option on physical property measurement system (PPMS-9, Quantum Design).

3. Results and discussions

3.1 Structural properties and phase purity

The XRD patterns of measurement and simulation were given in Fig.1. The experimental result revealed that the single-phase sample crystallizes in the ThCr2Si2-type body-centered tetragonal structure (shown in Fig.2a) with lattice constants, a = 4.115 and b = 9.593 Å, which are in good agreement with the ones reported previously [11]. For comparison, the lattice constants of isostructural compounds EuT_2P_2 (T = Fe, Co, Ni, and Cu) were plotted together in Fig.2, in which the data for the former three compounds are token from Refs [12]. The a value increases linearly when the transition metal changes from Fe to Cu due to the increase of radius of T. However, the dependence of c value on T is very different. The c increases linearly from $EuFe_2P_2$ to $EuCo_2P_2$, and deviates from this line for $EuNi_2P_2$ to $EuCu_2P_2$. They are about 25% reduced from the expected values on the line. It is attributed to the non-stoichiometry of the sample. In order to examine the composition, polished samples were checked by electron-probe microanalysis technique. The average result for nine checked points suggests a 13% deficiency of Cu.

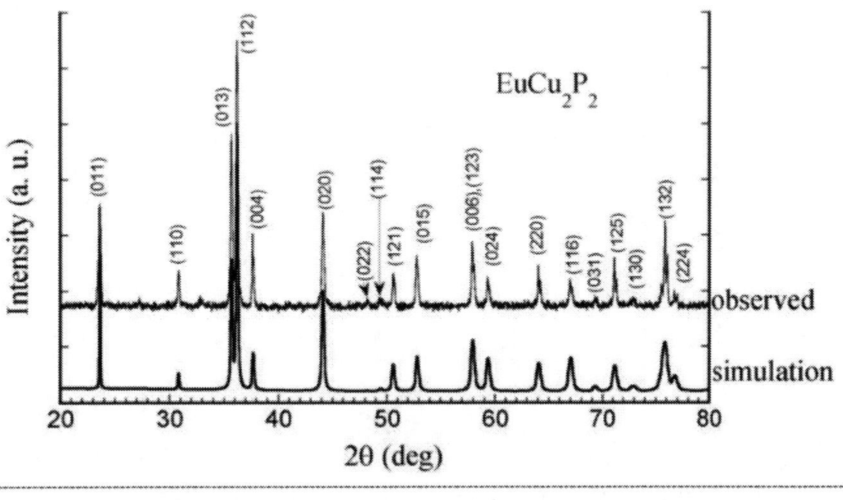

Figure 1. XRD patterns of $EuCu_2P_2$ indexed with$ThCr_2Si_2$-type structure.

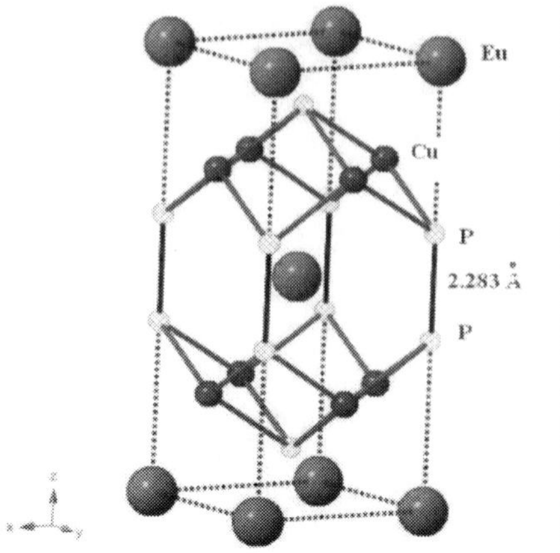

Figure 2a. Crystal structure of $EuCu_2P_2$.

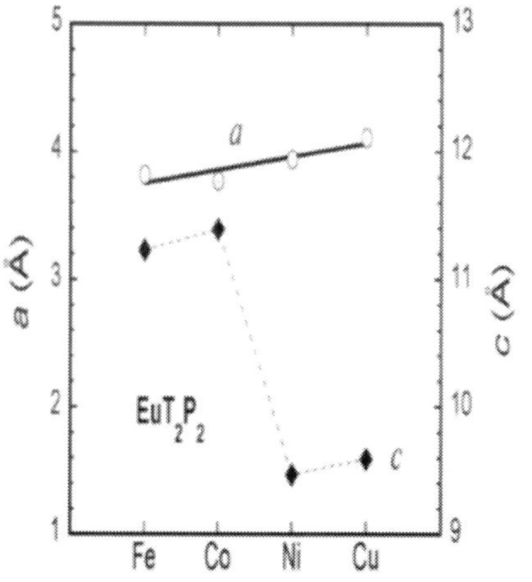

Figure 2b. The lattice constants *a* and *c* of EuT_2P_2.

3.2 Magnetic properties

The previous studies of the relationship between structure, bonding, and europium valence for this series suggests a correlation between the close phosphorus - phosphorus distance (d_{P-P}) along the *c* direction and the europium valence. The d_{P-P} decreases from a nonbonding distance greater than 3 Å in $EuFe_2P_2$ and $EuCo_2P_2$, in which europium is in divalent state [13,14], to a single bond distance of less than 2.4 Å for $EuNi_2P_2$, in which europium is in intermediate valence state [15]. The lattice constant of *c* for $EuNi_2P_2$ and $EuCu_2P_2$ are comparable, and single phosphorus - phosphorus bond (2.283 Å) was formed in both of them. However, the expected intermediate valence state in $EuCu_2P_2$ was excluded by the Eu Mössbauer and LIII-Edge X-Ray absorption spectroscopy [16]. Our magnetic measurement will give direct evidences of europium valence state and the information for magnetic order.

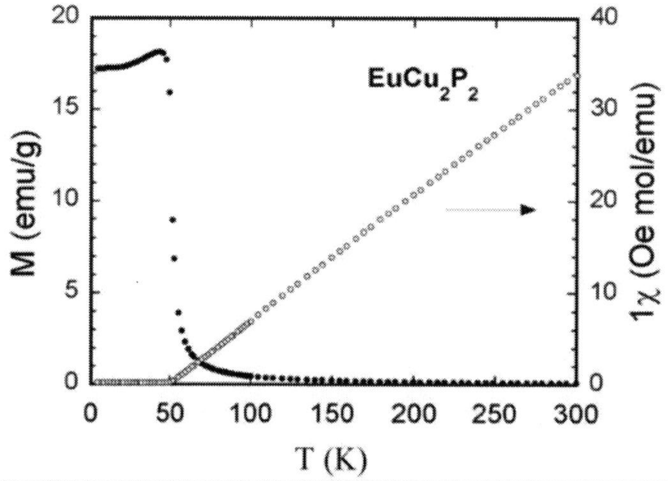

Figure 3. The temperature dependence of magnetization (M) and inverse susceptibility ($1/\chi$) for $EuCu_2P_2$.

The temperature dependence of magnetization (M) for $EuCu_2P_2$ in an external magnetic field H = 0.1 T is shown in Fig.3. The inverse susceptibility ($1/\chi$) is also plotted. A sharp increase was observed in the M-T curve with decreasing temperature, which suggesting a paramagnetic to ferromagnetic (PM-FM) transition. The Curie temperature T_C is about 51 K. In the paramagnetic region ($51\ K < T < 300\ K$), it shows the Curie-Weiss behaviour. The paramagnetic Curie temperature θ_P is obtained to be 46 K which is closed to the T_C. The estimated value of effective magnetic moment is 7.72 μ_B /Eu atom, which is closed to the free ion value of Eu^{2+} ($7.94\mu_B$). These results confirmed that the formation of a phosphorus - phosphorus bond alone is not sufficient to drive the transition to intermediate valence.

In fact, Pn-Pn bonding states in the EuT_2Pn_2 family strongly affect their electronic structure and magnetism. For EuT_2P_2 (T = Fe, Co, Ni, and Cu), the information for the magnetic structure was just reported for $EuCo_2P_2$ by neutron scattering study [14]. The Eu moments are ferromagnetic within the basal planes. Along the c-axis the order is that of an antiferromagnetic spiral. Although there are no magnetic moment of 3d sublattice, the 4f sublattice is ferromagnetically coupled in $EuFe_2P_2$ (T_C = 27 K) [13], and antiferromagnetically coupled in $EuCo_2P_2$ (T_N = 66 K)[14]. However, $EuNi_2P_2$ is a mixed valence compound. As stated above, the 4f sublattice ferromagnetic ordering occurs in $EuCu_2P_2$ again (T_C = 51 K). The difference in the magnetic order of Eu 4f moments in this series compounds could be understood in terms of an indirect RKKY interactions. The RKKY exchange coupling coefficient J_{RKKY} is tuned to be positive for T = Fe, negative for T = Co, nearly zero for T = Ni, and positive for T = Cu by the distance between magnetic Eu ions.

3.3 Electrical resistivity

The temperature dependence of resistivity (ρ) for $EuCu_2P_2$ was shown in Fig.4. The room temperature resistivity is 134 $\mu\Omega$ cm, which is comparable to that of polycrystalline sample of $EuFe_2P_2$ [8]. It decreases nearly linearly with decreasing temperature from room temperature to 51 K, and decreases drastically at lower temperatures. The kink around 50 K corresponds to the ferromagnetic transition in Eu sublattice as observed in magnetic measurement. The residual resistivity ratio (RRR) is smaller than that of $EuFe_2P_2$. [8] This is attributed to the atomic disorder due to the Cu deficiency.

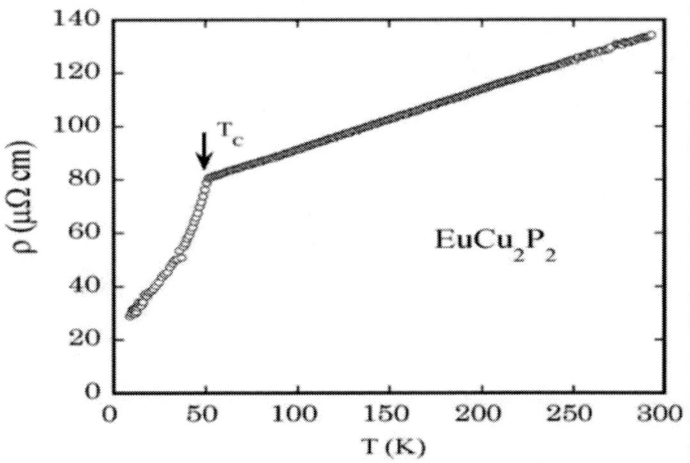

Figure 4. The temperature dependence of resistivity (ρ) for $EuCu_2P_2$.

4 Conclusions

In summary, the divalent state of europium in $EuCu_2P_2$ is confirmed by our magnetic and transport properties measurements. A ferromagnetic ordering of the Eu^{2+} below 51 K was observed. The magnetic interaction in the EuT_2P_2 is very sensitive to the distance between the magnetic ions and the P-P distance between the T_2P_2 layers.

Acknowledgements This work was partially supported by the National Natural Science Foundation of China (No. 50871036, 11004044), Major Scientific and Technological Innovation Project for Higher Education of Zhejiang Province (No. ZD2007010), and Innovation Research Team for Spintronic Materials and Devices of Zhejiang Province.

References

[1] A. Szytula and J. Leciejewicz , *Handbook of Crystal Structures and Magnetic Properties of Rare Earth Intermetallics* (CRC Press, Boca Raton, FL, 1994).

[2] F. Steglich, J. Aarts, C.D. Bredl, W. Lieke, D. Meschede, W. Franz, H. Sehafer, *Phys. Rev. Lett.* **43** (1977) 1892.

[3] K. Yoshimura, T. Nitta, M. Mekata, T. Shimizu, T. Sakakibara, T. Goto and G. Kido, *Phys. Rev. Lett.* **60** (1988) 851.

[4] G. Wortmann, I. Nowik, B. Perscheid and G. Kaindl, *Phys. Rev. B* **43** (1991) 5261.

[5] N. D. Mathur, F. M. Grosche, S. R. Julian, I. R. Walker, D. M. Freye, R. K. W. Haselwimmer, and G. G. Lonzarich, *Nature* (London) **394** (1998) 39.

[6] S. Raymond and D. Jaccard, *Phys. Rev. B* **61** (2000) 8679.

[7] Z. Ren, Z.W. Zhu, S. Jiang, X. F. Xu, Q. Tao, C. Wang, C. M. Feng, G. H. Cao and Z. A. Xu, *Phys. Rev. B* **78** (2008) 052501.

[8] Z. Ren, Q. Tao, J. Shuai, C. M. Feng, C. Wang, J. H. Dai, G. H. Cao and Z. A. Xu, *Phys. Rev. Lett.* **102** (2009) 137002.

[9] B. Ni, M. M. Abd-Elmeguid, and H. Micklitz, J. P. Sanchez and P. Vulliet, and D. Johrendt, *Phys. Rev. B* **63**, (2001) 100102.

[10] C. M. Feng, Z. Ren, S.G. Xu, Z. A. Xu, G. H. Cao, I. Nowik, I.Felner, K. Matsubayashi, and Y.Uwatok, arXiv:1005.0516v1 [cond-mat.str-el] 4 May 2010.

[11] A. Mewis, *Z. Naturforsch*, **35b** (1980)141.

[12] R. Marchand and W. Jeitschko, *J. Solid State Chem.* **24** (1978) 351.

[13] E. Morsen, B. D. Mosel, W. Muller-Warmuth, M. Reehuis, and W. Jeitschko, *J. Phys. Chem. Solids* **49** (1988) 785.

[14] M. Reehuis, W. Jeitschko, M. H. Möller, and P. J. Brown, *J. Phys. Chem. Solids* **53**, 687 (1992).

[15] S. Danzenbacher, D.V. Vyalikh, Yu. Kucherenko, A. Kade, C. Laubschat, N. Caroca-Canales, C.Krellner, C.Geibel, A.V. Fedorov, D. S. Dessau, R. Follath, W. Eberhardt, and S. L. Molodtsov, *Phys. Rev. Lett.* **102** (2009) 026403.

[16] R. M. Bornick and A. M. Stacy, *J. Solid State Chem.* **144** (1999) 252.

Journal of Physics: Conference Series **263** (2011) 012015 doi:10.1088/1742-6596/263/1/012015

Giant magnetoelectric effects in multilayered composites of cobalt ferrite and lead magnesium niobate titanate

Xiaohong Zhu[1], Hui Yang, Dingquan Xiao and Jianguo Zhu

Department of Materials Science, Sichuan University, Chengdu 610064, China

Email: xhzhu@scu.edu.cn

Abstract. The magnetoelectric (ME) coupling responses of $[CoFe_2O_4/(1-x)Pb(Mg_{1/3}Nb_{2/3})O_3-xPbTiO_3]_n$, represented as $(CFO/PMNT)_n$, multilayered composites were theoretically investigated by using an average method. It is found that a stronger interface coupling could be obtained in multilayers than in bilayers if the number of layers (n) is reasonably increased. Using an empirical dependence of the interface coupling parameter on the number of layers n, it is theoretically estimated that the ME coupling effect is first enhanced and then deteriorated with the gradual increase in n. The best ME response is estimated to occur at n=15, where the transverse ME voltage coefficient $\alpha_{E,31}$ reaches as high as 658 $mV \cdot cm^{-1} \cdot Oe^{-1}$. Moreover, it is theoretically found that the ME coupling effect in $(CFO/PMNT)_n$ multilayered composites is strongly dependent on the volume fraction of the piezoelectric (or magnetostrictive) phase, and the ME voltage coefficient for transverse field orientation ($\alpha_{E,31}$) is estimated to be roughly 90% higher than that for the longitudinal case ($\alpha_{E,33}$), revealing a large anisotropy in the ME coupling effect of $(CFO/PMNT)_n$ multilayered composites. These findings pave the way for experimental achievement of strong ME coupling responses, and provide important implications for the design and performance optimization of related devices comprising this kind of multiferroic magnetoelectric materials.

1. Introduction

Multiferroic magnetoelectric (ME) materials, which simultaneously exhibit ferroelectricity and ferromagnetism, have recently become the focus of research because of their scientific interest and significant technological promise in the novel multifunctional devices. Natural multiferroic single-phase compounds are rare, and their magnetoelectric responses are either relatively weak or occur at temperatures too low for practical applications. In contrast, multiferroic composites, which incorporate both ferroelectric and ferri-/ferromagnetic phases, typically yield giant magnetoelectric coupling responses above room temperature, making them ready for technological applications, such as applications in magnetic field sensors, transducers, magnetoelectric memories and microwave devices (filters, oscillators, phase shifters, etc) [1,2]. The magnetoelectric effect is defined as the dielectric polarization of a material in an applied magnetic field or an induced magnetization in an external electric field. That is, the induced polarization **P** is related to the magnetic field **H** by the expression, $\mathbf{P}=\alpha\mathbf{H}$, where α is the second rank ME-susceptibility tensor. A parameter of importance is the ME voltage coefficient $\alpha_E=\delta E/\delta H$ with $\alpha=\varepsilon_0\varepsilon_r\alpha_E$. A multilayer structure is expected to be far superior to bulk composites since the leakage current problem can be avoided and the piezoelectric layer can

[1] To whom any correspondence should be addressed.

easily be poled electrically to enhance the piezoelectricity and the ME effect [3]. In particular, multilayered magnetoelectric composites with cobalt ferrite ($CoFe_2O_4$, abbreviated as CFO) have attracted great interest due to the very high magnetostriction in CFO [4-7]. Moreover, the material system of lead magnesium niobate titanate [$(1\text{-}x)Pb(Mg_{1/3}Nb_{2/3})O_3\text{-}xPbTiO_3$, abbreviated as PMNT] is known to possess very high piezoelectric coefficients with composition near the morphotropic phase boundary (x: 0.3-0.4) [8-10]. As a consequence, the combination of highly magnetostrictive CFO and highly piezoelectric PMNT is a good choice for studying the magnetoelectric effect in multilayered composites. Until now, however, very few results have been reported in literature for multilayered composites composed of CFO and PMNT.

In this work, the magnetoelectric coupling responses of $(CFO/PMNT)_n$ multilayered composites were theoretically investigated by using an average method derived by Bichurin et al. [11,12]. In this model, an interface coupling parameter k that describes the actual boundary conditions is introduced. It is theoretically demonstrated that a stronger interface coupling could be obtained in multilayers than in bilayers if the number of layers is reasonably increased; moreover, the ME coupling effect in $(CFO/PMNT)_n$ multilayered composites is strongly dependent on the volume fraction of the piezoelectric (or magnetostrictive) phase, and there is a large anisotropy in the ME coupling effect of $(CFO/PMNT)_n$ multilayered composites.

2. Theoretical model

An averaging method was used for deriving the effective material parameters of CFO-PMNT composites and was performed in two stages [11,12]. In the first stage, the composite is considered as a bilayer consisting of magnetostrictive CFO and piezoelectric PMNT phases. It is noted that we consider only symmetric extensional deformation in this model and ignore any asymmetric flexural deformations of the layers. The magnetostrictive phase with a cubic symmetry can be expressed by the following equations:

$$\begin{aligned} {}^{m}S_i &= {}^{m}s_{ij}\,{}^{m}T_j + {}^{m}q_{ki}\,{}^{m}H_k, \\ {}^{m}B_k &= {}^{m}q_{ki}\,{}^{m}T_i + {}^{m}\mu_{kn}\,{}^{m}H_n, \end{aligned} \tag{1}$$

where ${}^{m}S_i$ and ${}^{m}T_j$ are strain and stress tensor components of the magnetostrictive phase, ${}^{m}H_k$ and ${}^{m}B_k$ are the vector components of magnetic field and magnetic induction, ${}^{m}s_{ij}$ and ${}^{m}q_{ki}$ are compliance and piezomagnetic coefficients, and ${}^{m}\mu_{kn}$ is the magnetic permeability matrix. For the polarized piezoelectric phase with the symmetry ∞m, the strain and electric displacement can be written as

$$\begin{aligned} {}^{p}S_i &= {}^{p}s_{ij}\,{}^{p}T_j + {}^{p}d_{ki}\,{}^{p}E_k, \\ {}^{p}D_k &= {}^{p}d_{ki}\,{}^{p}T_i + {}^{p}\varepsilon_{kn}\,{}^{p}E_n, \end{aligned} \tag{2}$$

where ${}^{p}S_i$ and ${}^{p}T_j$ are strain and stress tensor components of the piezoelectric phase, ${}^{p}E_k$ and ${}^{p}D_k$ are the vector components of electric field and electric displacement, ${}^{p}s_{ij}$ and ${}^{p}d_{ki}$ are compliance and piezoelectric coefficients, and ${}^{p}\varepsilon_{kn}$ is the dielectric permittivity matrix. Then, the ME voltage coefficients can be obtained from Eqs. (1) and (2) assuming in-plane mechanical connectivity between the two phases. In order to describe the actual boundary conditions, an interface coupling parameter (k) is introduced, which takes the form

$$k = ({}^{p}S_i - {}^{p}S_{i0})/({}^{m}S_i - {}^{p}S_{i0}), \tag{3}$$

where ${}^{p}S_{i0}$ is strain tensor components with no friction between layers. The k factor depends on the interface quality and is a measure of differential deformation between piezoelectric and magnetostrictive layers. The coupling factor k=1 is for an ideal interface and zero for the case with no friction. In the second stage, the bilayer is considered as homogeneous and the behaviour is described by

$$\begin{aligned} S_i &= s_{ij}T_j + d_{ki}E_k + q_{ki}H_k, \\ D_k &= d_{ki}T_i + \varepsilon_{kn}E_n + \alpha_{kn}H_n, \\ B_k &= q_{ki}T_i + \alpha_{kn}E_n + \mu_{kn}H_n, \end{aligned} \tag{4}$$

where α_{kn} is the ME coefficient. Effective parameters of the composite are obtained by solving Eq. (4), taking into account the solutions of Eqs. (1) and (2), for which the electric and magnetic vectors are determined using open- and closed-circuit conditions.

For detailed description, a coordinate system is assumed with the sample in the (1,2) plane. The sample is poled with an electric field E along the direction 3. ME coupling is estimated from the induced field δE across the sample that is subjected to an ac magnetic field δH in the presence of a bias field H. Basic relations for ME coefficients are obtained for two orientations of H and δH: along direction 3 (longitudinal to δE) or along direction 1 (transverse to δE), as depicted in Fig. 1. Therefore, the longitudinal ME voltage coefficient is given by

$$\alpha_{E,33} = \delta E_3/\delta H_3 = \frac{-2\mu_0 kv(1-v)\,{}^p d_{31}\,{}^m q_{31}}{2({}^p d_{31})^2(1-v)k + {}^p\varepsilon_{33}[({}^p s_{11} + {}^p s_{12})(v-1) - kv({}^m s_{11} + {}^m s_{12})]} \times \frac{[({}^p s_{11} + {}^p s_{12})(v-1) - kv({}^m s_{11} + {}^m s_{12})]}{[\mu_0(v-1) - {}^m\mu_{33}v][kv({}^m s_{11} + {}^m s_{12}) - ({}^p s_{11} + {}^p s_{12})(v-1)] + 2({}^m q_{31})^2 kv^2}, \tag{5}$$

where $v = {}^p v/({}^p v + {}^m v)$, representing the volume fraction of piezoelectric phase; ${}^p v$ and ${}^m v$ denote the volume of piezoelectric phase and magnetostrictive phase, respectively. The transverse ME voltage coefficient is given by

$$\alpha_{E,31} = \delta E_3/\delta H_1 = \frac{-kv(v-1)\,{}^p d_{31}({}^m q_{11} + {}^m q_{12})}{({}^m s_{11} + {}^m s_{12})\,{}^p\varepsilon_{33}kv + ({}^p s_{11} + {}^p s_{12})\,{}^p\varepsilon_{33}(1-v) - 2({}^p d_{31})^2 k(1-v)}. \tag{6}$$

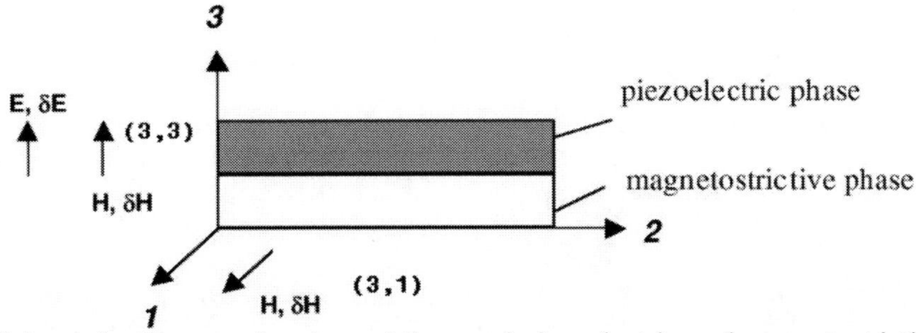

Figure 1. Schematic diagram showing a bilayer of piezoelectric and magnetostrictive phases in the (1,2) plane. Field orientations for longitudinal (3,3) and transverse (3,1) ME effects are also depicted.

The difference between ME coefficients for bilayer and multilayer is due primarily to an imperfect bonding of layers. As a result, the aforementioned k factor should be strongly dependent on the number of layers n in multilayered composites. Using an empirical formula extracted from experimental data for $NiFe_2O_4$-$Pb(Zr,Ti)O_3$ composites [13,14]:

$$k = 1 - (n-15)^2/225, \tag{7}$$

we can estimate the dependence of ME coupling effect on the number of layers in $(CFO/PMNT)_n$ multilayered composites.

3. Results and discussion

Let us now use the theoretical model for the calculation of ME coupling in $(CFO/PMNT)_n$ multilayered composites. The following sample parameters were used: for PMNT, the composition was selected as $0.68Pb(Mg_{1/3}Nb_{2/3})O_3$-$0.32PbTiO_3$, then s_{11}=47.7×10^{-12} m^2/N, s_{12}=-13.1×10^{-12} m^2/N, d_{31}=-686×10^{-12} C/N and $\varepsilon_{33}/\varepsilon_0$=4970 [15]; for CFO, s_{11}=6.48×10^{-12} m^2/N, s_{12}=-2.73×10^{-12} m^2/N, q_{11}=-1880×10^{-12} m/A, q_{12}=556×10^{-12} m/A, q_{31}=556×10^{-12} m/A and μ_{33}/μ_0=2 [16].

Since the strength of the ME coupling depends sensitively on the concentration of the two phases, we estimate the ME voltage coefficients as a function of the piezoelectric volume fraction v. Figure 2 shows the calculated dependence of longitudinal ($\alpha_{E,33}$) and transverse ($\alpha_{E,31}$) ME coefficients on v for $(CFO/PMNT)_n$ multilayered composites with a series of k values. Both $\alpha_{E,33}$ and $\alpha_{E,31}$ coefficients first increase and then decrease with the increase in the piezoelectric volume fraction v, reaching a maximum at v_m. It can also be seen from the figure that the decrease in k value leads to a significant lowering of the strength of ME coupling. This is because the ME coupling weakens with decreasing k. As a consequence, with increasing k, both the maximum $\alpha_{E,33}$ and $\alpha_{E,31}$ display a trend of near-linear increase. The ME coupling, however, reaches a maximum at progressively decreasing v_m with the increase in k. In other words, as k value increases, v_m shifts to CFO-rich compositions. These results are summarized and plotted in Fig. 3. It is noteworthy that a giant ME coupling effect is predicted here in $(CFO/PMNT)_n$ multilayered composites: the maximum $\alpha_{E,33}$ and $\alpha_{E,31}$ reach as high as 347 and 658 mV·cm^{-1}·Oe^{-1}, respectively, for an ideal interface coupling k=1, which are a little higher than those obtained in CFO-$Pb(Zr,Ti)O_3$ composites [11] and are probably attributed to very strong piezoelectric coupling for PMNT. Also important to note is that the ME voltage coefficient for transverse field orientation ($\alpha_{E,31}$) is estimated to be roughly 90% higher (nearly double) than that for the longitudinal case ($\alpha_{E,33}$), revealing a large anisotropy in the ME coupling effect of $(CFO/PMNT)_n$ multilayered composites. One could attribute this to a stronger parallel piezomagnetic coupling q_{11} (which determines $\alpha_{E,31}$) compared to q_{31} (which determines $\alpha_{E,33}$) [12].

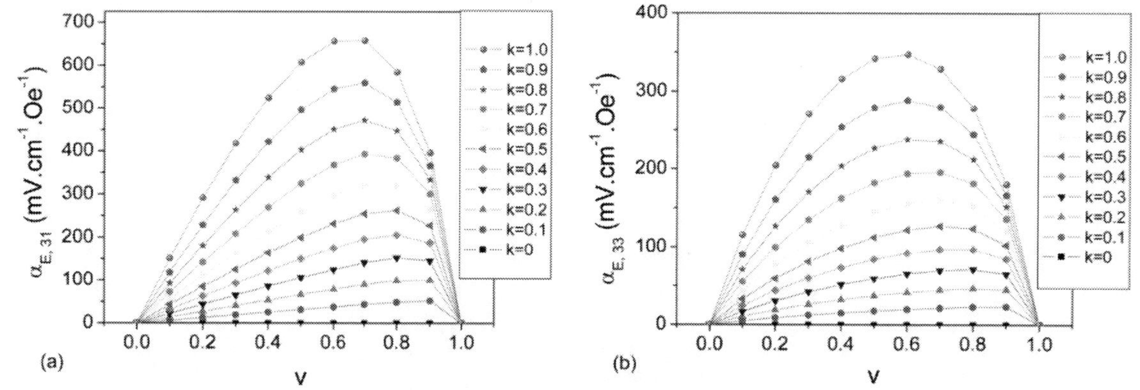

Figure 2. Dependence of the transverse ($\alpha_{E,31}$) and longitudinal ($\alpha_{E,33}$) magnetoelectric voltage coefficients on the volume fraction v of the piezoelectric phase in $(CFO/PMNT)_n$ multilayered composites. Theoretical estimates are shown as a function of the interface coupling parameter k.

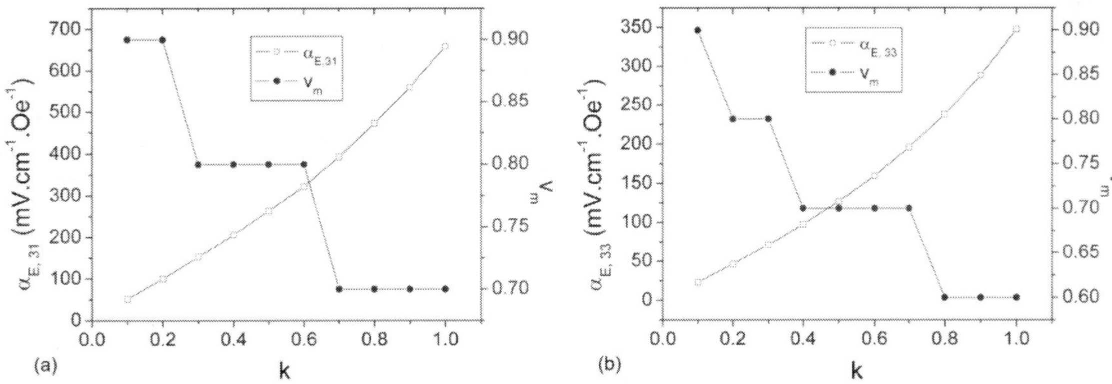

Figure 3. Variation with k of maximum $\alpha_{E,31}$ (a) and $\alpha_{E,33}$ (b) and the corresponding v_m for $(CFO/PMNT)_n$ multilayered composites.

It is shown in Fig. 4 that the transverse and longitudinal ME voltage coefficients depend drastically on the number of layers in $(CFO/PMNT)_n$ multilayered composites. For a given v value, the maximum ME coupling is obtained at n=15 for both $\alpha_{E,33}$ and $\alpha_{E,31}$, where the strongest interface coupling is achieved. This implies that the number of layers should be optimized in multilayered magnetoelectric composites in order to obtain a strong interface coupling and thus a strong ME coupling effect. Similarly, it is experimentally verified that the ME coupling effect could be enhanced in multilayered composites if the number of layers is reasonably increased [14,17,18]. It is pointed out that the volume fraction of the two phases also has a remarkable effect on the ME coupling, especially for the longitudinal ME response. In the case of $(CFO/PMNT)_n$ multilayered composites, the maximum ME coupling occurs at n=15 and v=0.6. Hence, both the number of layers and the volume fraction of piezoelectric phase should be carefully taken into consideration in the design of high-performance multilayered magnetoelectric composites.

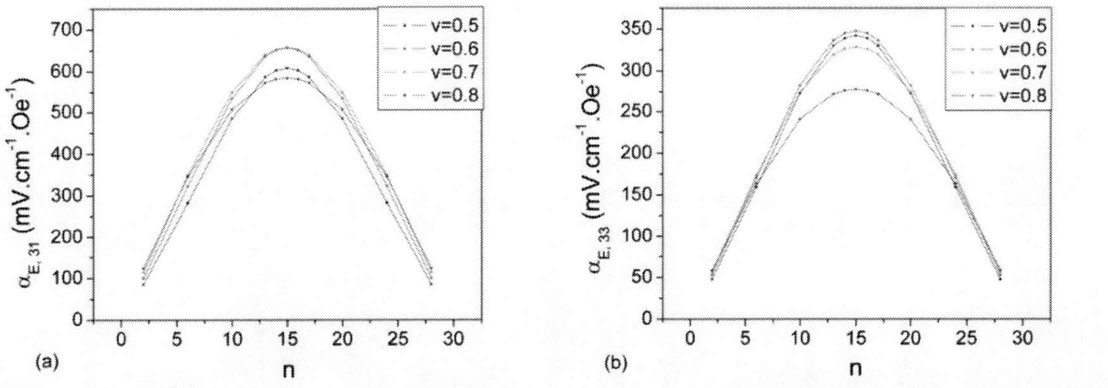

Figure 4. Peak transverse (a) and longitudinal (b) ME voltage coefficients as a function of the number of layers in $(CFO/PMNT)_n$ multilayered composites with different piezoelectric volume fractions.

4. Conclusions

The magnetoelectric coupling responses of $(CFO/PMNT)_n$ multilayered composites were theoretically investigated by using an average method. It is found that the ME coupling effect in $(CFO/PMNT)_n$ multilayered composites is strongly dependent on the volume fraction of the piezoelectric phase (v), and the ME voltage coefficient for transverse field orientation ($\alpha_{E,31}$) is estimated to be roughly 90%

higher (nearly double) than that for the longitudinal case ($\alpha_{E,33}$), revealing a large anisotropy in the ME coupling effect of $(CFO/PMNT)_n$ multilayered composites. It is also found that a stronger interface coupling and thus a stronger ME coupling effect could be obtained in multilayers than in bilayers if the number of layers (n) is reasonably increased. Using an empirical dependence of the interface coupling parameter on the number of layers, it is theoretically estimated that the ME coupling effect is first enhanced and then deteriorated with the gradual increase in n. The best ME response is estimated to occur at n=15 and ν=0.6, where the transverse ME voltage coefficient $\alpha_{E,31}$ reaches as high as 658 $mV \cdot cm^{-1} \cdot Oe^{-1}$. These findings pave the way for experimental achievement of strong ME coupling responses, and provide important implications for the design and performance optimization of related devices comprising this kind of multiferroic magnetoelectric materials.

Acknowledgements
This work was supported by the Program for New Century Excellent Talents in University of China (NCET-10-0582).

References

[1] Eerenstein W, Mathur N D and Scott J F 2006 *Nature* **442** 759
[2] Nan C W, Bichurin M I, Dong S X, Viehland D and Srinivasan G 2008 *J. Appl. Phys.* **103** 031101
[3] Srinivasan G, Rasmussen E T, Gallegos J, Srinivasan R, Bokhan Y I and Laletin V M 2001 *Phys. Rev. B* **64** 214408
[4] Wan J G, Wang X W, Wu Y J, Zeng M, Wang Y, Jiang H, Zhou W Q, Wang G H and Liu J M 2005 *Appl. Phys. Lett.* **86** 122501
[5] Zhou J P, He H C, Shi Z and Nan C W 2006 *Appl. Phys. Lett.* **88** 013111
[6] Lin R J, Wu T B and Chu Y H 2008 *Scripta Mater.* **59** 897
[7] Zhang J X, Dai J Y, Chow C K, Sun C L, Lo V C and Chan H L W 2008 *Appl. Phys. Lett.* **92** 022901
[8] Fu H X and Cohen R E 2000 *Nature* **403** 281
[9] Noheda B, Cox D E, Shirane G, Gao J and Ye Z G 2002 *Phys. Rev. B* **66** 054104
[10] He C J, Zhou D, Wang F F, Xu H Q, Lin D and Luo H S 2006 *J. Appl. Phys.* **100** 086107
[11] Bichurin M I, Petrov V M and Srinivasan G 2002 *J. Appl. Phys.* **92** 7681
[12] Bichurin M I, Petrov V M and Srinivasan G 2003 *Phys. Rev. B* **68** 054402
[13] Bichurin M I, Petrov V M and Srinivasan G 2002 *Ferroelectrics* **280** 165
[14] Srinivasan G, Hayes R, Devreugd C P, Laletsin V M and Paddubnaya N 2005 *Appl. Phys. A* **80** 891
[15] Lee S H and Roh Y 2007 *Jpn. J. Appl. Phys.* **46** 4462
[16] Wang L,Yang C T, Xie Q Z and Ye J H 2009 *Acta Phys. Sin.-Ch.Ed.* **58** 3515
[17] Zhang J X, Dai J Y and Chan H L W 2010 *J. Appl. Phys.* **107** 104105
[18] Liu Y, Zhang Y, Lin Y H and Nan C W 2009 *J. Appl. Phys.* **105** 083915

Electronic structures and the spin polarization of Heusler alloy Co_2FeAl surface

Xiaoguang Xu, Yankai Wang, Delin Zhang, Yong Jiang[1]

State Key Laboratory for Advanced Metals and Materials, School of Materials Science and Engineering, University of Science and Technology Beijing, Beijing 100083, China

E-mail: yjiang@ustb.edu.cn

Abstract. The electronic structures of the Heusler alloy Co_2FeAl surface are studied theoretically via first-principles calculations. The space localization of the surface states is the key effect on the electronic structures of the Co_2FeAl surface. At the surface, the lattice parameter shrinks to minimize the total energy, and the minority spin gap disappears and shows a metallic band gap character. However, with the depth increasing, the lattice parameter equals to that of bulk phase, and there shows an energy gap opening at the Fermi level in the minority spin states. As a result, the spin polarization at the surface is lower than that of the bulk Co_2FeAl, while it is close to that of bulk phase beneath the surface. According to the calculations, it is clear that the half-metallic property fading of the Co_2FeAl films is caused by the surface states. Therefore, it is important to minimize the lattice mismatch at the interface of Co_2FeAl in order to obtain a high tunneling magnetoresistance.

1. Introduction

Half-metallic ferromagnets (HMFs), which are metallic in the majority spin while insulating in the minority spin, have a theoretical spin polarization of 100%. Among all the HMFs, Co-based full-Heusler alloys are ideal materials for application as ferromagnets in spintronic devices, such as spin valves (SPVs) or magnetic tunnel junctions (MTJs), because of their high Curie temperature (T_c) and low coercivity (H_c). In recent years, Co-based Heusler alloys have attracted great interest. The application in MTJs as ferromagnetic layers can dramatically improve tunneling magnetoresistance (TMR). Okamura et al. [1] fabricated $Co_2FeAl/Al\text{-}O_x/Co_{75}Fe_{25}$ MTJ on a thermally oxidized Si substrate, and obtained a TMR of 47% at room temperature (RT). Co_2FeSi is also reported to be a HMF with high T_c and magnetic moment, both experimentally and theoretically. [2, 3] The Fermi level lies at the top of the minority spin gap in Co_2FeSi, while at the bottom in Co_2FeAl. [4] So theoretically, the $Co_2FeAl_{1-x}Si_x$ ($0<x<1$) compound has perfect half-metallic electronic structures with the Fermi level locating in the middle of the minority spin gap. [5] It is also experimentally verified that the application of $Co_2FeAl_{1-x}Si_x$ in MTJs can improve TMR dramatically. [6] However, the TMR value varies in a wide range, even with the same HMFs electrodes and MgO barrier. [7, 8] The MTJs with HMFs electrodes are theoretically predicted to have an infinite TMR value. But till now, the highest TMR value realized experimentally is only a few hundred percent. This is due to different

[1] To whom any correspondence should be addressed.

surface or interface structures that are essential to the performance of MTJs. Therefore, it is important to study the surface electronic structures of HMFs.

2. Calculations

Our density functional theory calculations are performed by using Vienna *ab initio* Simulation Package (VASP). [9] Exchange correlation interactions are described by the Perdew-Bruke-Ernzerholf (PBE) generalized gradient approximation (GGA). [10] The projector-augmented wave (PAW) method [11] is used for the treatment of core electrons. In addition, the GGA+U method is used to consider on-site correlation of 3*d* transition metals. The effective Coulomb-exchange interaction $U_{eff} = U - J$ is used, where U and J are the Coulomb part and exchange part, respectively. The electronic structures are sensitive to the value of U_{eff}, and a large U_{eff} results in a large band gap. Therefore, we select smaller values of U_{eff} to avoid overestimated band gap in the calculations (U_{Co} = 2.2 eV, U_{Fe} = 2.1 eV). In order to test the reliability of the calculations, we first calculate the electronic structures of bulk Co_2FeAl in $L2_1$ phase. The positions of Al, Fe and Co are (0, 0, 0), (1/2, 1/2, 1/2) and (1/4, 1/4, 1/4), respectively. We employ an 8x8x8 mesh of special *k*-points to perform the summation over the Brillouin zone. The energy convergence criterion is set to 10^{-3} eV and the criterion for charge convergence to 10^{-5} electrons. The lattice parameter is 5.692 Å after optimization, which is close to the experimental value 5.730 Å. [12] The supercells to model the $Co_2FeAl(001)$ surface are constructed by 3, 4, and 5 periodic unit cell of Co_2FeAl $L2_1$ phase separated by a vacuum layer of 12 Å, which is named as 3, 4, and 5 layers, respectively. The model of 3 layers is presented in Figure 1. The lattice parameters are fixed, and the atoms are relaxed using the same convergence criterion as that of the bulk phase. A *k*-mesh of 6x6x1 is employed for geometry optimizations, while a 6x6x6 *k*-mesh for density of states (DOS) calculations. The kinetic cutoff energy is 335 eV in all calculations.

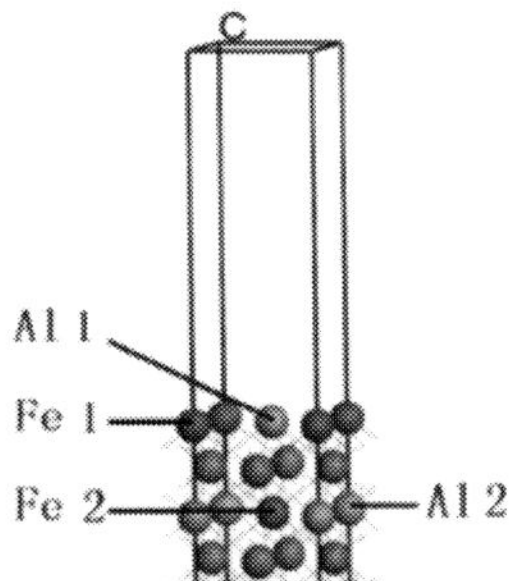

Figure 1. Supercell of the $Co_2FeAl(001)$ surface.

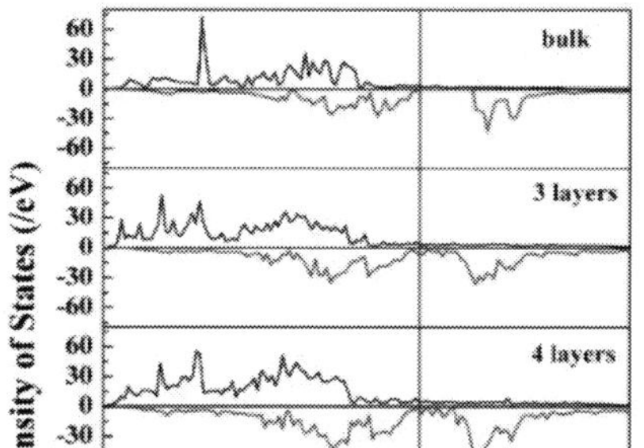

Figure 2. Ttotal DOS for the Co_2FeAl bulk phase and surface.

3. Results and discussion

The total DOS of the Co_2FeAl bulk phase and the (001) surface of 3, 4, and 5 layers are presented in Figure 2, in which the zero of energy denotes the Fermi level. The majority states of the bulk Co_2FeAl take on a metallic character, while the minority states have an energy gap about 1.0 eV with the Fermi level locating at the bottom of it. The electronic structures indicate that the bulk Co_2FeAl in $L2_1$ phase is half-metallic, which is in consistent with the result of G. H. Fecher et al..[4] For the total DOS of the $Co_2FeAl(001)$ surface, it can be seen in all the three panels that both the majority and minority states are metallic, which is the main difference between the bulk phase and the (001) surface causing the fading of half-metallic property.

To justify whether the calculation model is thick enough to describe the surface, we calculate the total energy and magnetic moment for each model. The total energy difference between 3 and 4 layers is 45.2 eV, which is close to the value of 45.3 eV between 4 and 5 layers. The energy fluctuation of 0.0125 eV per atom is within the calculation error. The magnetic moment increments from 3 to 4 layers and from 4 to 5 layers are 10.8 and 10.6 μ_B, respectively. The changes of energy and magnetic moment with the surface thickness are almost constant, which suggests the 3 layers model is thick enough to discuss the surface effect. Accordingly, we choose the 3 layers model to study detailed DOS.

In Figure 3, we present total DOS and partial density of states (PDOS) of Al, Fe, and Co atoms in the 3 layers model. The atoms at different depth as shown in Fig. 1 are sampled to reveal the evolutional electronic structures with the depth beneath the surface. From the total DOS shown in the first row of Fig. 3, it can be seen that both the majority and minority states are metallic for the atoms locating at the positions numbered as '1' in Fig. 1 (represented by solid lines in Fig. 3), which are exposed to vacuum at the surface of Co_2FeAl. When the depth increases and the atoms locate at the positions numbered as '2' (represented by dashed lines in Fig. 3), the majority states are still metallic and have no distinct change at the Fermi level, while the minority states of each atom decrease dramatically relative to those of the atoms in the positions '1'. The atoms at the positions '3' (marked in Fig. 1) become deeper beneath the surface, and are separated with vacuum by five atomic layers, which is deep enough to be regarded as in the bulk environment. The total DOS of Al, Fe, and Co atoms in the positions '3' all show metallic character in the majority states. However, in the minority states of each atom, there opens an energy gap at the Fermi level, indicating an insulating character. According to the evolution of total DOS, it is found that in the ordered $Co_2FeAl(001)$ surface, the spin polarization increases with the atom depth beneath the surface. When the atom depth is close to 0.8 nm which is deeper than the positions '3', the spin polarization equals to that of the bulk phase.

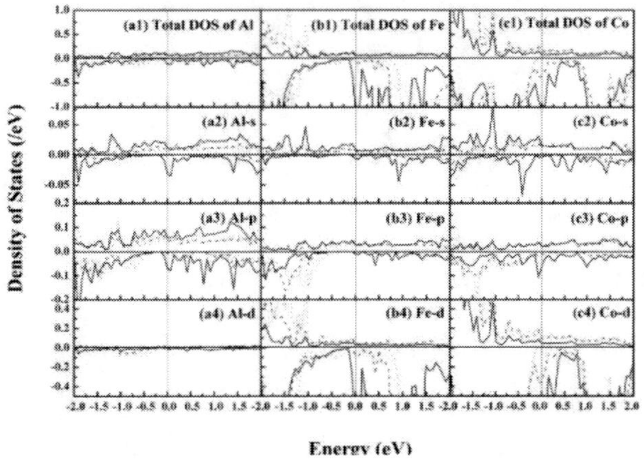

Figure 3. PDOS of atoms at different depth in 3 layers surface.

The change of spin polarization with the atom depth beneath the $Co_2FeAl(001)$ surface comes from the distribution of PDOS, which is sensitive to the coordinate environment. According to the geometry optimization results, the bond length at the $Co_2FeAl(001)$ surface will shrink to minimize the total energy. In the optimized surface, the distance between the first layer and the second one at the FeAl termination is 1.278 Å, from which the lattice parameter in z direction at the surface is calculated to be 5.112 Å. While in the middle of the surface, the optimized lattice parameter is close to that of the bulk phase. Similar trend can also be observed in the Co termination. Accordingly, the bond length increases with the depth beneath the surface and finally reaches the value of the bulk phase, which is accompanied by the evolution of electronic orbits. As shown in Fig. 3 column (a), the s and p states are main contribution to the total DOS of Al. When Al depth changes from the positions of '1' to '3', the minority states of s and p decrease to zero at the Fermi level, so that there opens an energy gap. The PDOS patterns of Fe atoms are presented in Fig. 3 column (b). The minority states at the Fermi level mainly come from 3d states which decide the spin polarization. It is obvious that the 3d minority states at the Fermi level decreases dramatically with the Fe atom depth, and the 3d minority spin band at the Fermi level shifts to the high energy region. As shown in Fig. 3 column (c), the evolution of Co PDOS is similar with that of Fe, except that the Co 3d minority spin band at the Fermi level shifts to the low energy region with the Co atom depth increasing. The bond length at the surface is shorter than that beneath the surface, hence the electronic orbits largely overlap, and the energy levels are much degenerated. As a result, the 3d states disperse around the Fermi level due to the surface-localized state at the surface. However, beneath the surface, a longer bond length results in localized electronic states. The full occupied 3d majority states are not sensitive to the bond length increase, while in the partially occupied minority states, the band-gap opens up due to the decay of the surface-localized state beneath the surface. The minority charge density slices for (010) and (110) planes of the Co_2FeAl surface are shown in Figures 4(a) - (d), associated to the states within the energy region from -0.1 eV to 0.7 eV. Comparing the charge density of atoms at different depth, we can see that the minority state is metallic at the surface and insulated beneath the surface, which shows the space-localization states in the $Co_2FeAl(001)$ surface.

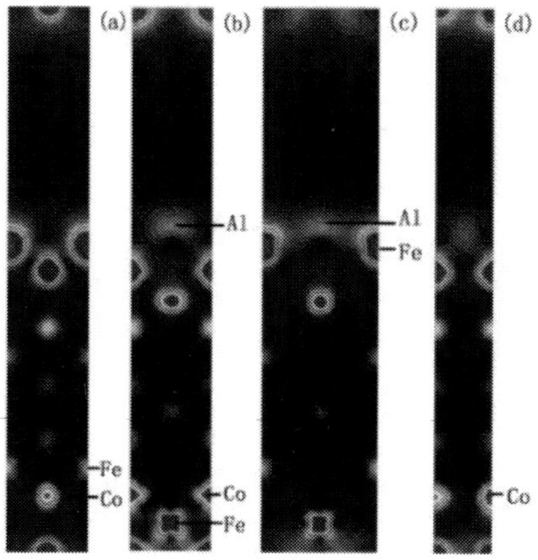

Figure 4. Minority charge density of Co_2FeAl surface.

The magnetic moments of Co and Fe atoms mainly come from the 3d states. The s and p states only have very small induced magnetic moments. It is found that the atomic magnetic moments of both Fe and Co atoms decrease with the depth beneath the surface. In the bulk Co_2FeAl $L2_1$ phase, the calculated magnetic moments of Fe and Co atoms are 3.0 and 1.2 μ_B, which are close to those in the positions '3' in Fig. 1. Therefore, the variation of the atomic magnetic moments also demonstrates that the magnetic properties become close to those of the bulk Co_2FeAl with the depth increasing. The saturation magnetization of the $Co_2FeAl(001)$ surface is higher than that of the bulk phase based on the simulation results.

According to the analysis above, the $Co_2FeAl(001)$ surface can be regarded as a composite tri-layer structure which has low spin polarization at the surface and high spin polarization beneath the surface. When Co_2FeAl is fabricated into thin films, the half-metallic property will fade due to the low spin polarization at the surface.

4. Conclusion

In summary, the first-principles calculations on the Co_2FeAl surface reveal that the minority states are metallic at the surface due to the crystal lattice distortion. However, the DOS of individual atoms show a trend of minority gap opening at the Fermi level beneath the surface, which indicates that the electronic structures become close to those of the bulk Co_2FeAl. Accordingly, the $Co_2FeAl(001)$ surface has a composite tri-layer structure, which can illustrate the half-metallic property fading in thin films against ideal character of bulk materials.

Acknowledgements
This work was partly supported by the NSFC (Grant Nos. 50701005, 50831002, 50971025), the Key Grant Project of the Chinese Ministry of Education (No. 309006), the National Basic Research Program of China (Grant No. 2007CB936202), and the Fundamental Research Funds for the Central Universities.

References
[1] Okamura S, Miyazaki A, Sugimoto S, Tezuka N and Inomata K 2005 *Appl. Phys. Lett.* **86** 232503
[2] Wurmehl S, Fecher G H, Kandpal H C, Ksenofontov V, Felser C, Lin H J and Morais J 2005 *Phys. Rev. B* **72** 184434
[3] Inomata K, Okamura S, Miyazaki A, Kikuchi M, Tezuka N, Wojcik M and Jedryka E 2006 *J. Phys. D: Appl. Phys.* **39** 816
[4] Fecher G H and Felser C 2007 *J. Phys. D: Appl. Phys.* **40** 1582
[5] Nakatani T M, Rajanikanth A, Gercsi Z, Takahashi Y K, Inomata K and Hono K 2007 *J. Appl. Phys.* **102** 033916
[6] Wang W, Sukegawa H, Shan R and Inomata K 2008 *Appl. Phys. Lett.* **93** 122506
[7] Wang W, Sukegawa H, Shan R and Inomata K 2008 *Appl. Phys. Lett.* **93** 182504
[8] Tezuka N, Ikeda N, Mitsuhashi F and Sugimoto S 2009 *Appl. Phys. Lett.* **94** 162504
[9] Kresse G and Furthmuller J 1996 *Phys. Rev. B* **54** 11169
[10] Perdew J P, Burke K and Ernzerhof M 1996 *Phys. Rev. Lett.* **77** 3865
[11] Kresse G and Joubert D 1999 *Phys. Rev. B* **59** 1758
[12] Buschow K H J and van Engen P G 1981 *J. Magn. Magn. Mater.* **25** 90

A comparative study of the magnetic properties and magnetic entropy change in RCo_2B_2 (R = Pr, Nd and Gd) compounds

Lingwei Li[a,b,c,1], **Dexuan Huo**[a,b], **Zhenghong Qian**[b], **and Katsuhiko Nishimura**[c]
[a]Institute of Materials Physics, Hangzhou Dianzi University, Hangzhou 310018, China
[b]Center for Integrated Spintronic Devices(CISD), Hangzhou Dianzi University, Hangzhou 310018, China
[c]Graduate School of Science and Engineering, University of Toyama, Toyama 930-8555, Japan

E-mail: lingwei@hdu.edu.cn

Abstract. A series of single-phased RCo_2B_2 (R = Pr, Nd and Gd) compounds were prepared by an arc-melting method. The structure, magnetic properties and magnetiocaloric effect (MCE) were studied by the magnetization and heat capacity measurement. A paramagnetic to ferromagnetic (PM-FM) transition was observed and the Curie temperature T_C is 18 and 27 K for $PrCo_2B_2$ and $NdCo_2B_2$, respectively. However, a paramagnetic to antiferromagnetic (PM-AFM) magnetic transition was observed ~15 K under low fields (H $\leqslant$ 0.2 T), whereas a field-induced metamagnetic transition from AFM to FM has been observed with the application of higher magnetic field (H $\geqslant$ 0.4 T) for $GdCo_2B_2$. A large/giant reversible MCE was observed around its magnetic transition temperature for RCo_2B_2 systems. For a field change of 5 T, the maximum values of magnetic entropy change ($-\Delta S_M^{max}$) reach 8.1, 7.1 and 17.1 J kg^{-1} K^{-1} for R = Pr, Nd and Gd, respectively. The magnetic transition, origin of large MCE and its potential application in RCo_2B_2 systems were discussed.

1. Introduction

In last two decades, the magnetic materials with large magnetocaloric effect (MCE) have been extensively studied experimentally and theoretically due to its great potential applications for magnetic refrigeration [1-20]. The MCE was first discovered in pure Iron by Emil Warburg in 1881 [1]. The origin of the magnetocaloric effect was independently explained later, in the 1920s, by Debye and Giauque [2,3]. At the time they proposed the adiabatic demagnetization in order to achieve temperature lower than that of the liquid helium. This possibility was later experimentally demonstrated by the chemist Nobel Laureate William F. Giauque in 1933, when with his equipment, based on the adiabatic demagnetization of paramagnetic salts, he was able to reach a temperature of 0.25 K [4]. Between 1933 and 1997, a number of advances in utilization of the MCE for cooling occurred. In 1997, two key developments enhanced the feasibility for producing a magnetic refrigerator for commercial or industrial use. The first is that Zimm et al [9] realized the first near room temperature magnetic refrigeration prototype at the Ames Laboratory. Second, in the same year, a giant MCE was discovered in the $Gd_5Si_2Ge_2$ compound by Pecharsky and Gschneider [10]. Since the discovery of the giant MCE in $Gd_5Si_2Ge_2$, lots of materials with large magnetic entropy changes have

[1] To whom any correspondence should be addressed.

been observed such as RCo_2 (R = Er, Ho, and Dy) [11, 12], $La(Fe_{1-x}Si_x)_{13}$ and its hydration [13, 14], $MnFeP_xAs_{1-x}$ [15, 16], and manganites $R_{1-x}M_xMnO_3$, (where R = lanthanide and M = Ca, Sr, and Ba) etc [17, 18].

Rare-earth intermetallic compounds of the RT_2X_2 have been extensively studied due to their interesting physical properties. Depending on the constituent element or composition, various properties like superconductivity, magnetic ordering, heavy-fermion properties, valence fluctuation and large MCE, etc. are observed [21-26]. Recently, a giant MCE in antiferromagnetic $ErRu_2Si_2$ [27] and $GdCo_2B_2$ [28] compounds (belong to RT_2X_2 series compounds) was observed, which was associated with a field – induced first order magnetic transition from AFM to FM state below the (zero field) AFM transition temperature. Soon after that the MCE in $PrCo_2B_2$ [29] and $NdCo_2B_2$ [30] was reported, however, the MCE was obviously smaller than that in $GdCo_2B_2$. To further understand the origin of the differences of MCE between $PrCo_2B_2$, $NdCo_2B_2$ and $GdCo_2B_2$ with the same crystal structure and find a clue for searching new materials with excellent MCE properties, a comparative study of the magnetic properties and magnetic entropy change in RCo_2B_2 (R = Pr, Nd and Gd) compounds are presented in this paper.

2. Experimental

The polycrystalline RCo_2B_2 (R = Pr, Nd, Gd and Y) were synthesised by an arc melting method using a tungsten electrode under an argon atmosphere. Stoichiometric amounts of high purity Pr, Nd, Gd, Y, Co and B were melted more than six times for homogeneity on a water-cooled copper hearth. The total weight loss of the sample in this step was less than 0.5%. Finally, the samples were annealed in evacuated quartz tubes at 900 °C for 1 week. The phase composition and structure were characterized by X-ray powder diffraction experimental (XRD). The magnetization measurements were done using a superconducting quantum interference device magnetometer (Quantum Design, MPMS-7). The specific heat measurements were carried out by the adiabatic heat relaxation method using a physical property measurement system (PPMS-9) from Quantum Design. The results were well repeatable.

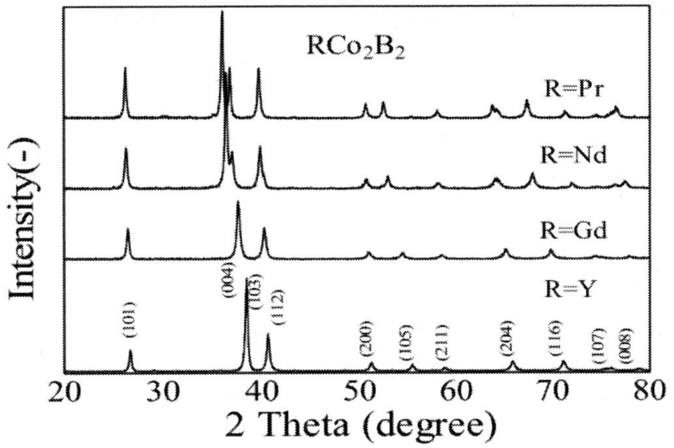

Figure 1. X-ray diffraction patterns for RCo_2B_2 (R = Pr, Gd, Nd and Y).

3. Results and discussion

The X-ray diffraction (XRD) measurements were performed on all of the samples; the results are shown in Fig. 1. No impurity peaks were found within experimental errors; all the samples are single phased with the body centered tetragonal crystal belonging to I/4mmm space group. The lattice parameters a and c were evaluated from XRD data using Rietveld refinement method, and the results are shown in Table 1. The results are consistent with those of previously reported results [31, 32]. Figure 2 shows the temperature dependence of magnetization (M) in an external magnetic field H =

0.1 T for RCo_2B_2 (R = Pr, Nd and Gd). A sharp paramagnetic to ferromagnetic (PM-FM) transition can be observed and the Curie temperature T_C is 18 and 27 K (defined as the inflection point of M-T curve) for $PrCo_2B_2$ and $NdCo_2B_2$, respectively. However, a clear paramagnetic to antiferromagnetic (PM-AFM) magnetic transition can be observed and the Neel temperature T_N is around 15 K for $GdCo_2B_2$. The transition temperatures were consistent with those previously reported by Rupp et al. [31]. The magnetization as a function of temperature in various external magnetic fields, H = 0.1 - 1 T, for $GdCo_2B_2$ are shown in the inset of Fig. 2. The nature of M (T) under 0.1 and 0.2 T indicates AFM ordering below ~15 K, whereas a field-induced metamagnetic transition from AFM to FM has been observed with the application of higher magnetic field (H ≥ 0.4 T).

Table 1. Lattice parameter a and c for RCo_2B_2 (R = Pr, Nd, Gd and Y).

Sample	a (Å)	c (Å)	a (Å) Ref.[1]	c (Å)Ref. [1]
$PrCo_2B_2$	3.600	9.946	3.599	9.951
$NdCo_2B_2$	3.592	9.846	3.592	9.838
$GdCo_2B_2$	3.577	9.537	3.574	9.555
YCo_2B_2	3.562	9.346	3.561	9.353

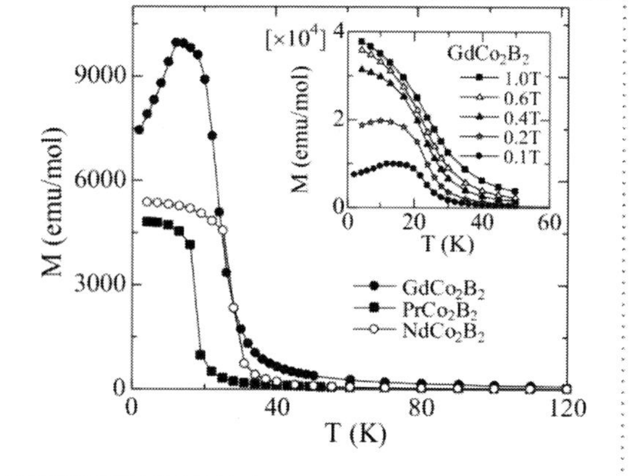

Figure 2. Temperature dependence of magnetization (M, H = 0.1 T) for RCo_2B_2 (R = Pr, Gd, Nd and Y). Inset: The magnetization as a function of temperature for $GdCo_2B_2$ in various external magnetic fields, H = 0.1 - 1 T.

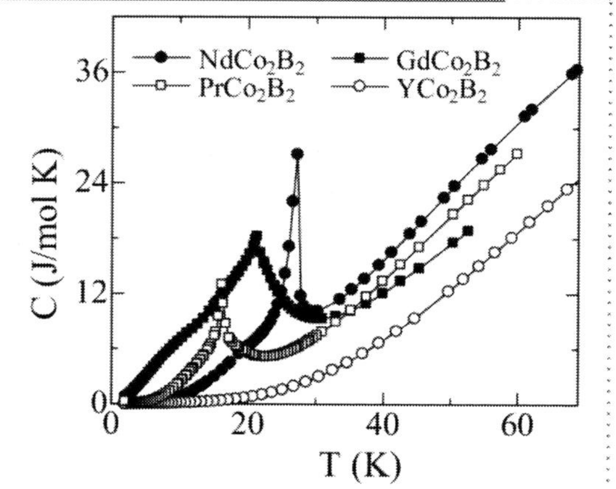

Figure 3. Temperature dependence of specific heat results for RCo_2B_2 (R = Pr, Nd, Gd and Y) under zero applied magnetic field.

The specific heat results for RCo_2B_2 (R = Pr, Nd, Gd and Y) in zero applied magnetic field are shown in Fig. 3. These data provide a determination for T_M = 16, 27 and 21 K, respectively for R = Pr, Nd and Gd, which are consistent with those deduced from $M(T)$ curves. To derive the magnetic contribution C_{mag} from the total specific heat actually measured, C_{tot}, the following expression is commonly used:

$$C_{tot} = C_{lat} + C_{mag} = C_{el} + C_{ph} + C_{mag}. \quad (1)$$

C_{lat} represents the lattice contribution to the specific heat, including the conduction electron part C_{el} and the phonon part C_{ph}. The lattice contributions can be estimated from the measurement of specific heat of isomorphous nonmagnetic compound; in the present study, that of YCo_2B_2, which has the same crystal structure and without magnetic ordering, was used. The renormalization of Debye temperature due to the different molar masses between Y and R was carried out following the many-Debye method [33], respectively. In this model, for a ternary compound $R_mX_nY_p$ the renormalization ratio γ_{re} could be determined by

$$\gamma_{re} = \frac{\theta_D(R_m X_n Y_p)}{\theta_D(R'_m X_n Y_p)} = \left[\frac{m(M_{R'})^{3/2} + n(M_X)^{3/2} + p(M_Y)^{3/2}}{m(M_R)^{3/2} + n(M_X)^{3/2} + p(M_Y)^{3/2}} \right]^{1/3}, \quad (2)$$

where M_R, M_X, M_Y being the molar mass of R, X and Y atoms, respectively. The lattice contribution to heat capacity $(RCo_2B_2)_{lat}$, calculated from the specific heat results of YCo_2B_2 with the renormalization ratio, γ_{re} = 0.8815, 0.8750 and 0.8506 respectively for R = Pr, Nd and Gd. The difference between the C_{tot} and C_{lat} of RCo_2B_2 gives a measure of magnetic heat capacity C_{mag}. The calculated results are shown in Fig. 4. The sharp peak near its own T_M is due to the magnetic phase transition. The additional broad peak around 42 and 56 K respectively for R = Pr and Nd is related to the Schottky-anomaly [29, 30]. Using the formula

$$S_M = \int_0^T \frac{C_{mag}}{T} dT, \quad (3)$$

the temperature dependence of the magnetic entropy S_M is calculated (as shown in Fig. 5). S_M gradually increases with increasing temperature. At the transition temperature T_C, the value of S_M is close to Rln2 for $PrCo_2B_2$ and $NdCo_2B_2$, which means that the ground state of Pr and Nd ion forms a doublet state in RCo_2B_2. For $GdCo_2B_2$, S_M saturated with a value close to Rln8 above T_M, which means that the total spin of Gd ion in the ground state of $GdCo_2B_2$ is 7/2.

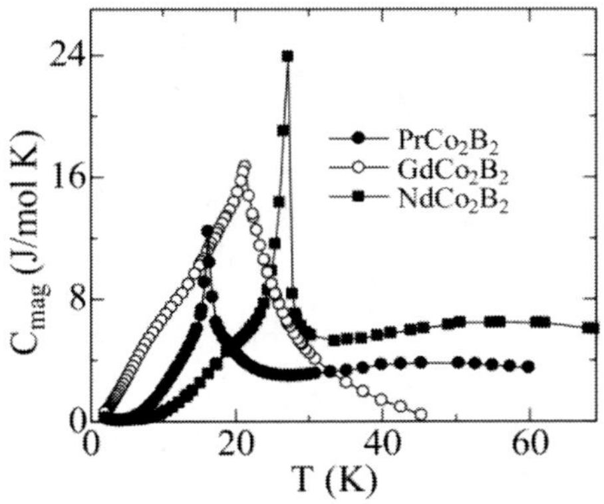

Figure 4. Temperature dependence of magnetic heat capacity C_{mag} for RCo_2B_2 (R = Pr, Nd, and Gd) .

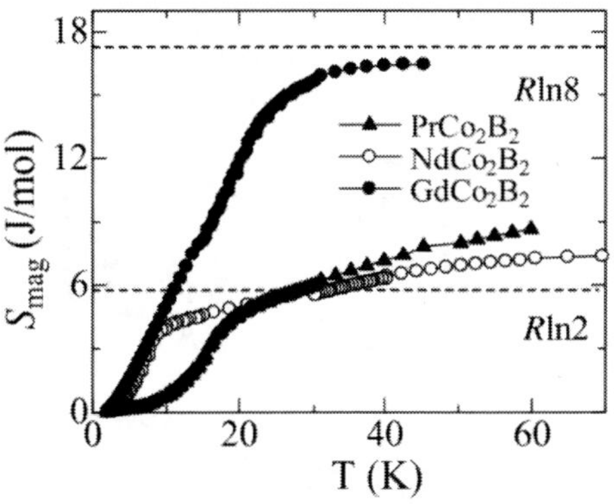

Figure 5. Temperature dependence of Zero field magnetic entropy S_{mag} for RCo_2B_2 (R = Pr, Nd, and Gd).

A set of magnetic isothermals on increasing and decreasing field were measured for RCo_2B_2 (R = Pr, Nd and Gd) with the temperature range from 4 to 62 K up to 7 T. There is no obvious hysteresis for $PrCo_2B_2$ and $NdCo_2B_2$, whereas, for $GdCo_2B_2$ the hysteresis and metamagnetic transition can be observed at T = 4 and 10 K. (see inset of Fig. 6 (c)). For low temperature ones, the magnetization M tends to be saturated at low fields. A large reversible MCE is expected around the transition temperature where the magnetization rapidly changes with varying temperature. The Inoue-Shimizu model, which involves a Landau expansion of the magnetic free energy up to sixth power of the total magnetization M, can be used to determine the order of magnetic transition [34],

$$F(M,T) = \frac{c_1(T)}{2} M^2 + \frac{c_2(T)}{4} M^4 + \frac{c_3(T)}{6} M^6 + \cdots - BM. \quad (4)$$

The parameters $c_1(T)$, $c_2(T)$ and $c_3(T)$ represent the Laudau coefficient, and it has been reported that the order of a magnetic transition is related to the sign of the $c_2(T)$. A transition is expected to be the first order when $c_2(T_C)$ is negative, whereas it will be the second order for a positive $c_2(T_C)$. The sign of $c_2(T_C)$ can be determined by means of Arrott plots [35, 36]. If the Arrott plot is S-shaped near T_C,

$c_2(T_C)$ is negative, otherwise, positive. To further understand the nature of the magnetic transition in RCo$_2$B$_2$, the Arrott-plots M^2 vs. H/M at some selected temperatures for RCo$_2$B$_2$ (R = Pr, Nd and Gd) are plotted in Fig. 6. Neither the inflection point nor negative slopes is observed, providing the occurrence of a second order magnetic transition for PrCo$_2$B$_2$ and NdCo$_2$B$_2$. GdCo$_2$B$_2$ undergoes a first order magnetic phase transition because there is a clear S-shaped behavior below the transition temperature.

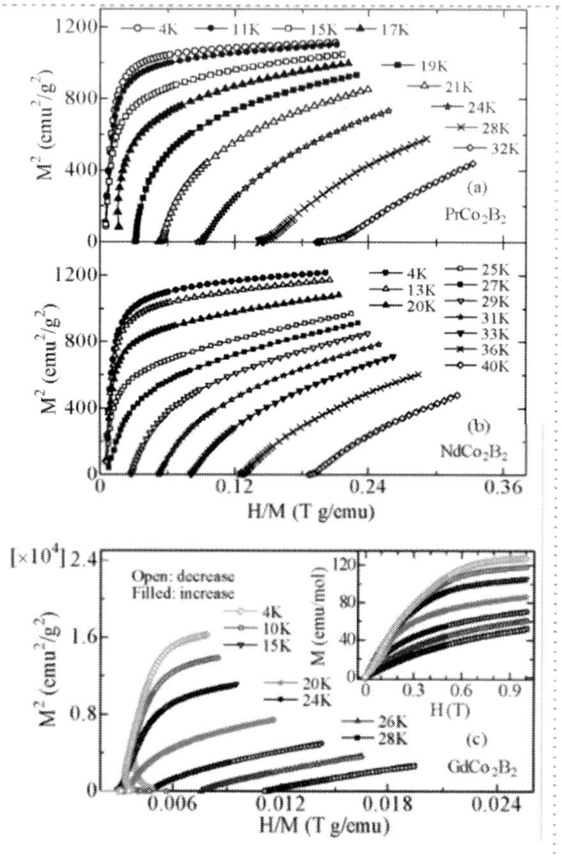

Figure 6. (Colour online) The Arrot-plot of RCo$_2$B$_2$ (R = Pr, Nd and Gd) at some selected temperatures. The inset of (c) shows low field magnetic isothermals on increasing (open symbols) and decreasing (filled symbols) field at some selected temperatures for GdCo$_2$B$_2$.

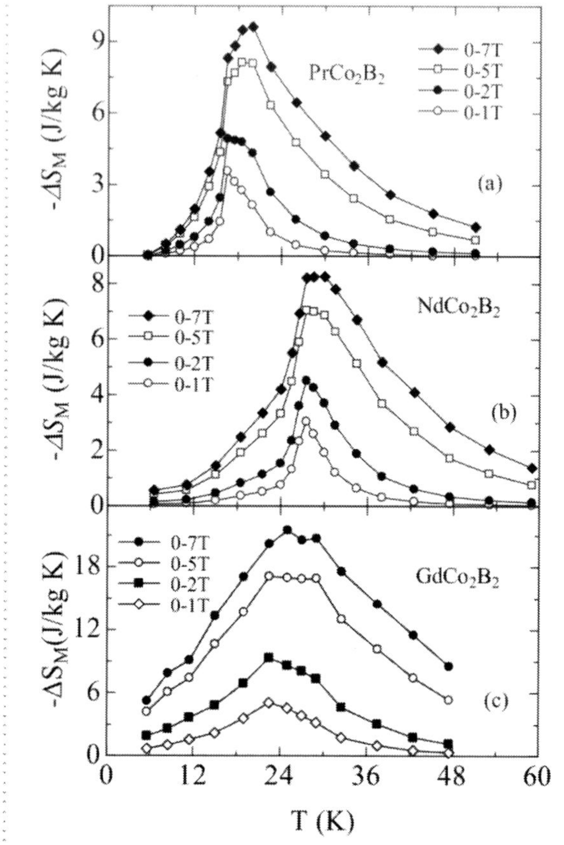

Figure 7. The magnetic entropy change -ΔS_M as a function of temperature for various magnetic field changes up to 7 T for RCo$_2$B$_2$ (R = Pr, Nd and Gd).

The magnetic entropy change -ΔS_M was calculated from magnetization isotherm curves using an integral version of Maxwell's thermodynamic relation [37]:

$$\Delta S_M(T,\Delta H) = \int_0^H \left(\frac{\partial M(H,T)}{\partial T}\right)_H dH \quad . \qquad (5)$$

The resulting changes of magnetic entropy ΔS_M as a function of temperature for different magnetic field variations up to 7 T are shown in Fig. 7 (a), (b) and (c), respectively for PrCo$_2$B$_2$, NdCo$_2$B$_2$ and GdCo$_2$B$_2$. For field changes of 5 and 7T, the maximum values of magnetic entropy change (-ΔS_M^{max}) reach 8.1 and 9.6 J kg^{-1} K^{-1} for PrCo$_2$B$_2$ which were comparable with those of NdCo$_2$B$_2$ (7.1 and 8.3 J

kg^{-1} K^{-1}). This large MCE is believed to be related to the second order magnetic phase transition from PM to FM state. The values of $-\Delta S_M^{max}$ for $GdCo_2B_2$ reach 17.1 and 21.5 J kg^{-1} K^{-1} for the field changes of 5 and 7 T, respectively. Such giant MCE was believed to be related to the field induced first-order magnetic transition from AFM to FM, similar to those observed in Ho_5Pd_2 [38], DySb [39] and $ErRu_2Si_2$ [27] compounds.

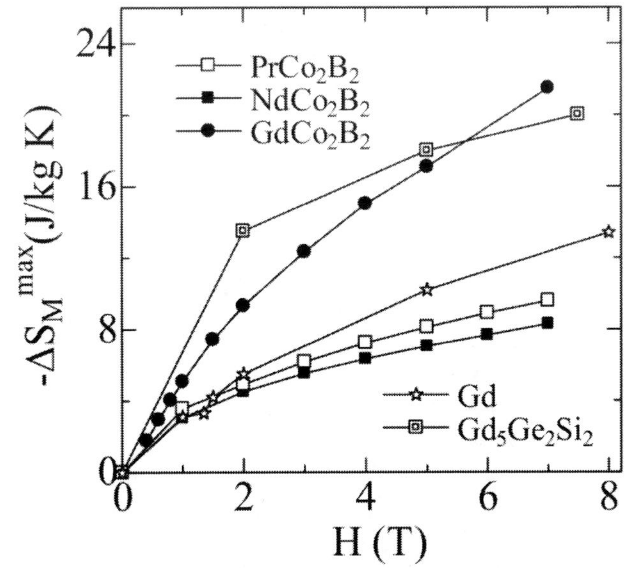

Figure 8. Field dependence of maximum magnetic entropy change $-\Delta S_M^{max}$ for RCo_2B_2 (R = Pr, Nd and Gd).

Table 2. The transition temperature (T_M), order of transition type, maximum values of magnetocaloric parameters ($-\Delta S_M^{max}$ and RCP) under 2 and 5 T for RCo_2B_2 (R = Pr, Nd and Gd) and some MCE materials.

Material	T_M (K)	Order of transition	$-\Delta S_M^{max}$ (J /kg K) 2T	5T	RCP (J /cm^3) 2T	5T	Ref.
$PrCo_2B_2$	16	second	4.9	8.1	0.36	1.1	present
$NdCo_2B_2$	27	second	4.5	7.1	0.3	0.8	present
$GdCo_2B_2$	25	first	9.3	17.1	1.06	3.8	present
$ErRu_2Si_2$	5.5	first	11	17.6	~0.8	~2.6	27
DySb	11	first	-	15.8	-	~1.4	39
$GdPd_2Si$	17	second	4.5	15	~0.4	~2.3	40
PrNi	20	second	2.4	6.1	~0.2	~0.7	41
Ho_5Pd_2	28	first	~7	~18	1.7	6.3	38
$ErCo_2$	35	first	28	33	1.5	3.08	11,19

The field dependence of $-\Delta S_M^{max}$ for RCo_2B_2 (R = Pr, Nd and Gd) as well as a few well known giant MCE materials, Gd and $Gd_5Si_2Ge_2$ [10, 19] are shown in the Fig. 8 for a comparison. The $-\Delta S_M^{max}$ increases linearly with the increasing field for $PrCo_2B_2$ and $NdCo_2B_2$ similar to that of $GdCo_2B_2$, but the values are obviously smaller than those of $GdCo_2B_2$. The transition temperature (T_M), the order of magnetic transition, the maximum values of magnetic entropy change ($-\Delta S_M^{max}$) and relative cooling power (RCP, evaluated by $RCP = -\Delta S_M^{max} \times \delta T_{FWHM}$) under 2 and 5T for RCo_2B_2 (R =

Pr, Nd and Gd) and various giant MCE materials are listed in Table 2 for a comparison. For $GdCo_2B_2$, the MCE parameters are comparable with some of the best known giant MCE materials around room temperature, and even larger than those of the potential promising magnetic refrigerant materials in the similar temperature region. The values of $-\Delta S_M^{max}$ for $PrCo_2B_2$ and $NdCo_2B_2$ are smaller than those of which processed the first order magnetic transitions, the present values are comparable with those of $GdPd_2Si$ [40] and PrNi [41] which undergo the second order magnetic transitions. Therefore, the differences of MCE between $PrCo_2B_2$, $NdCo_2B_2$ and $GdCo_2B_2$ with the same crystal structure could be understood if we recall the molar magnetic entropy which has a theoretical upper limit given as $S_M = R_A \ln(2J+1)$ J/mol K, where R_A is a universal gas constant and J is the total angular momentum. And the magnetocaloric effect should be larger at a transition which involves the large portion of the available magnetic entropy which was strongly dependent on the transition type (first order or second order). I.e. The larger MCE in $GdCo_2B_2$ than those of $PrCo_2B_2$ and $NdCo_2B_2$ were related to its prior magnetic transition (first order) and its higher molar magnetic entropy.

4. Conclusions

In summary, a comparative study of the magnetic properties and magnetic entropy change in RCo_2B_2 (R = Pr, Nd and Gd) compounds are presented. A PM-FM transition was observed for $PrCo_2B_2$ and $NdCo_2B_2$. However, A PM-AFM magnetic transition was observed under low fields, whereas a field-induced metamagnetic transition from AFM to FM has been observed with the application of higher magnetic field. A giant/large magnetic entropy change was observed around its transition temperature. For a field changes of 7 T, the maximum values of magnetic entropy change ($-\Delta S_M^{max}$) reaches 21.5 J kg^{-1} K^{-1} for $GdCo_2B_2$ which were obviously larger than those of $PrCo_2B_2$ (9.6 J kg^{-1} K^{-1}) and $NdCo_2B_2$ (8.3 J kg^{-1} K^{-1}). The giant MCE in $GdCo_2B_2$ was believed to be related to the field induced first-order metamagnetic transition from AFM to FM state different from those of in $PrCo_2B_2$ and $NdCo_2B_2$ (second-order magnetic transition from PM to FM state). The reasons for the larger MCE in $GdCo_2B_2$ than those of $PrCo_2B_2$ and $NdCo_2B_2$ are (i) it has larger molar magnetic entropy S_M and (ii) the transition thermodynamically reveals a first order magnetic phase transition. The present results could give us a clue to search new materials with excellent MCE properties.

5. Acknowledgements

This work was partially supported by the National Natural Science Foundation of China (No. 11004044, 50871036), Major Scientific and Technological Innovation Project for Higher Education of Zhejiang Province (No. ZD2007010), and Innovation Research Team for Spintronic Materials and Devices of Zhejiang Province.

6. References

[1] E. Warburg, Ann. Phys. (Leipzig) 13 (1881) 141.
[2] P. Debye, Ann. Phys. 81 (1926) 1154.
[3] W. F. Giauque, J. Am. Chem. Soc. 49 (1927) 1864.
[4] W. F. Giauque and D. P. MacDougall, Phys. Rev. 43 (1933) 768.
[5] V. K. Pecharsky and K. A. Gschneidner Jr, J. Magn. Magn. Mater. 200 (1999) 44.
[6] K. A. Gschneidner Jr and V. K. Pecharsky, Annu. Rev. Mater. Sci. 30 (2000) 387.
[7] A.M. Tishin, in: K.H. Buschow (Ed.), Handbook of Magnetic Materials, vol. 12, North-Holland, Amsterdam, 1999, pp. 395–524.
[8] E. Bruck, J. Phys. D: Appl. Phys. 38 (2005) R381.
[9] C. Zimm, A. Jastrab, A. Sternberg, V. K. Pecharsky, K. Gschneidner Jr, M. Osborne, I. Anderson, Adv. Cryog. Eng. 43 (1998) 1759.
[10] V. K. Pecharsky, and K. A. Gschneidner, Jr., Phys. Rev. Lett. 78 (1997) 4494.
[11] N. K. Singh, K. G. Suresh, A. K. Nigam, S. K. Malik, A. A. Coelho, S. Gama, J. Magn. Magn. Mater. 317 (2007) 68.
[12] Z. Gratz, and A. S. Markosyan, J. Phys.: Condens. Matter 13 (2001) R385.

[13] A. Fujita, S. Fujieda, Y. Hasegawa, Fukamichi, Phys. Rev. B 67 (2003) 104416.
[14] F. X. Hu, B. G. Shen, J. R. Sun, Z. H. Cheng, G. H. Rao, X. X. Zhang Appl. Phys. Lett. 78 (2001) 3675.
[15] O. Tegus, E. Bruck, K. H. J. Buschow, F. R. De Boer, Nature (London) 415 (2002) 150.
[16] N. Sun, S. Ma, Q. Zhang, J. Du, Z. Zhang, Appl. Phys. Lett. 91 (2007) 112503.
[17] L. Li, K. Nishimura, W. D. Hutchison, K. Mori, J. Phys. D : Appl. Phys. 41 (2008) 175002.
[18] M.-H. Phan, and S.-C. Yu, J. Magn. Magn. Mater. 308 (2007) 325.
[19] K. A. Gschneidner, Jr., V. K. Pecharsky, A. O. Tsoko, Rep. Prog. Phys. 68 (2005) 1479.
[20] V. K. Pecharsky, and K. A. Gschneidner, Jr., Int. J. Refrigeration 29, 1239 (2006).
[21] S. Di Napoli, A. M. Llois, G. Bihlmayer, S. Blügel, Phys. Rev. B 75 (2007) 104406.
[22] L. Li, K. Nishimura, Appl. Phys. Lett. 95 (2009) 132505.
[23] L. Li, K. Nishimura, K. Mori, Physica C 467 (2007) 9.
[24] S. Kervan, A. Kulic, S. Ozcan, A. Gencer, Solid State Commun. 130 (2004) 101.
[25] T. Zhang, Y. Chen, Y. Tang, E. Zhang, M. Tu. Phys. Lett. A 354 (2006) 462.
[26] G. Venturini, J. Alloys Compd. 232 (1996) 133.
[27] T. Samanta, I. Das, S. Banerjee, Appl. Phys. Lett. **91** (2007)152506.
[28] L. Li, K. Nishimura, H. Yamane, Appl. Phys. Lett. 94 (2009) 102509.
[29] L. Li and K. Nishimura, J. Appl. Phys., 106 (2009) 023903
[30] L. Li and K. Nishimura, J. Phys. D: Appl. Phy., 42 (2009) 145003.
[31] B. Rupp, P. Rogl and F. Hulliger, J. Less-Common Metals, 135 (1987) 113.
[32] I. Felner, Solid State Commun. 52 (1984) 191.
[33] M. Bouvier, P. Lethuillier, and D. Schmitt, Phys. Rev. B 43 (1991) 13137.
[34] J. Inoue and M. Shimizu, J. Phys. F: Met. Phys. 12 (1982) 1811.
[35] X. B. Liu and Z. Altounian, J. Magn. Magn. Mater. 292 (2005) 83.
[36] N. H. Duc, D. T. K. Anh and P. E. Brommer Physica B 320 (2002) 319.
[37] V. K. Pecharsky and K. A. Gschneidner Jr., J. Appl. Phys. **86** (1999)565.
[38] T. Samanta, I. Das, and S. Banerjee, Appl. Phys. Lett. **91** (2007) 082511.
[39] W. J. Hu, J. Du, B. Li, Q. Zhang, and Z. D. Zhang, Appl. Phys. Lett. **92** (2008) 192505.
[40] R. Rawat and I. Das J. Phys.: Condens. Matter 13 (2001) L57.
[41] A. O. Pecharsky, Yu. Mozharivskyj, K. W. Dennis, K. A. Gschneidner, Jr., R. W. McCallum, G. J. Miller, and V. K. Pecharsky, Phys. Rev. B 68 (2003) 134452.

$Zn_{1-x}Co_xO$ film valence by photoelectron spectroscopy

Yingzi Peng[1,2], **Dexuan Huo**[1] **and Liping Zhu**[2]

[1] Department of Physics, School of Science, Hangzhou Dianzi University, Hangzhou 310018, P.R.China

[2] State Key Lab of Silicon Materials, Zhejiang University, Hangzhou 310027, P.R.China

E-mail: yingzip@hdu.edu.cn

Abstract. The Co-doped ZnO film was fabricated using a pulsed laser deposition method on a C-sapphire substrate. The structural properties of the Co-doped ZnO film and the chemical states of elements have been studied in this paper. The XRD and HRTEM characterization results show that it is wurtzite structure with the C-axis of the film aligned with that of the substrate without obvious precipitation. Furthermore the experimental results probably reveal the occupation of Co ions at Zn sites in the host lattice. The oxygen ions are lying in metal oxide, which are due to O - Zn(Co) bonds, though a few oxygen deficiencies and oxygen vacancies were also detected. Co ions have replaced part of Zn sites in the host lattice, resulting in Co - O bonds and no obvious precipitate was detected in the film under our experimental conditions. The substituted Co ions are in the 2+ oxidation state. The electronic-structure parameters Δ, U are also induced.

1. Introduction

Diluted magnetic semiconductors (DMSs) are semiconductors with some magnetic ions as impurities in the host lattice. Generally, DMSs are synthesized by semiconductors doped with 3d transition metal ions while their semiconductor properties are maintained. The discoveries of InMnAs and GaMnAs in the 1990s re-activated research in magnetic semiconductors [1] and stimulated the development of a new field called spintronics. Since then, how to develop new DMSs with higher Curie temperatures became a common concern. Dietl theoretically predicted that the Curie temperatures of DMSs, such as ZnO and GaN based compounds, could be raised to room temperature within a mean-field picture [2]. After that, ZnO doped with 3*d* transition metals has become a hot topic for research [3-5]. Though many experimental results have been reported on Co-doped ZnO, there are controversies on the reports of ZnO-based DMSs studies, such as the inconsistency of results reported by different groups and confused explanations of the magnetic behaviors. Therefore, recently it has received much attention to explore the physical mechanisms underlying ZnO-based DMSs [6]. Among these, the studies on valence or chemical states of materials were attracted much interests [7-9]. For example, D. Seghier et al. studied a deep electron trap in ZnO and Co-doped ZnO epilayers and suggested a

[1] Corresponding author: Yingzi Peng, yinzip@hdu.edu.cn

correlation between the observed ferromagnetism in these materials and the oxygen vacancy [10]. Ref [11] reported the ferromagnetism ordering of the materials could be switched between "on" and "off" by introducing or removing, respectively, the oxygen vacancies in Co-doped ZnO.

In our previous papers, the observation of room temperature *M-H* loops of $Zn_{1-x}Co_xO$ films were already reported [12-14]. This work is following our previous reports on ZnO based diluted magnetic semiconductors, in which Co-doped ZnO thin films were synthesized using a dual-beam pulsed-laser-deposition method. This paper addresses chemical states of the films prepared using a single beam pulsed-laser-deposition (PLD) method by photoelectron spectroscopy, which can provide useful chemical-state information and relatively straightforward interpretation of spectra. In this paper, very sophisticated computer curve fitting routines will be used.

2. Experiment

The Co-doped ZnO film was fabricated by a pulsed laser deposition method from a ceramic target on a C-plane sapphire substrate. A pulsed KrF excimer laser (Lamba physic) was used as a light source. The laser was set to operate at a repetition rate of 10 Hz, laser energy around 110 mJ. A single crystal sapphire substrate was used to support the crystal growth of the thin film. The substrate was put on the substrate holder at a distance of 50 mm from the target. The film was formed at 650 °C in ambient pressure around 5×10^{-5}Torr. The target for PLD was prepared by a self-assembly ceramic technique using ZnO and Co powders following the procedure of mixing, compacting and calcining. The Co contents of targets are 5 at%.

The specimen was analyzed by X-ray diffraction (XRD) to identify the crystal structures and orientation of the films. X-ray photoelectron spectroscopy (XPS) was performed to determine element chemical states of the components. Detailed lattice structures and possible precipitates were obtained by a high-resolution transition electron microscopy (HRTEM).

3. Results and discussions

The Co-doped ZnO film was examined by XRD. Figure 1 shows the XRD pattern of the specimen on the sapphire substrate. The graph shows a clear XRD pattern, that is, besides the peak corresponding to the (0006) of substrate, only diffraction peaks corresponding to the (0002) and (0004) planes of the Co-doped ZnO film can be clearly observed. Note that presenting in the logarithm scale, besides the peaks corresponding to $\lambda_{Cu\alpha1}$, the peaks of (0002) for the film and the (0006) for the substrate corresponding to $\lambda_{Cu\alpha2}$ and $\lambda_{Cu\beta}$ were also detected, indicating high orientation growth of the film. No peak of other phases were detected. Consequently, it is a wurtzite structure with the C-axis of the film aligned with that of the substrate. The (0002) peak of the film is close to that of ZnO, probably implying that part of Zn sites in the host lattice were replaced by Co ions. The inset shows the rocking curve around (0002) peak of the specimen. The value of full-width at half maximum (FWHM) of the peaks is around 0.2°. Based on the results, the dispersion of the orientation is relatively small and the film is expected to grow along a preferred direction.

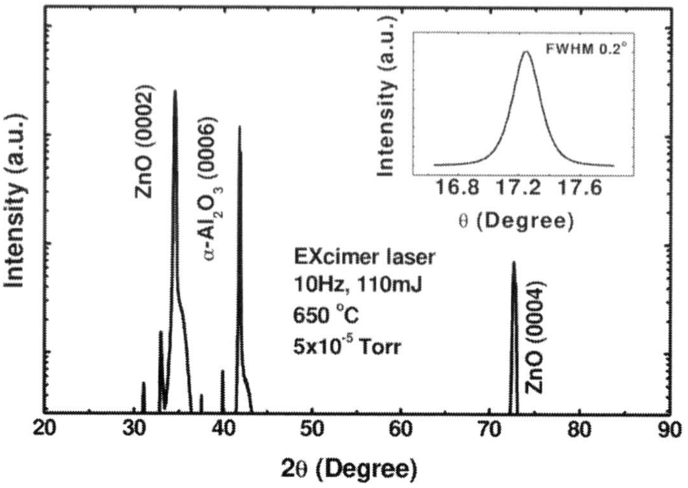

Figure 1. XRD pattern of the Co-doped ZnO film. The inset shows the rocking curve around (0002) peak of the specimen.

In order to study the details of the crystal structures, HRTEM of the Co-doped ZnO film was characterized. Figure 2 presents the HRTEM cross-section image and the selected area electron diffraction pattern of the film. The film lattice is shown in Figure 2(a), showing a very clear lattice. Figure 2(b) shows a selected area electron diffraction pattern of the film. It can be seen that it is a wurtzite structure without obvious precipitation. HRTEM for substrate and interface were also characterized (not shown here), showing that the C-axis of the film aligned with that of the substrate, which are similar to our previous results. These results seem to coincide with the results of XRD. According to our experimental results, no obvious precipitate was detected. These results probably reveal a suggestion of the occupation of Co ions at Zn sites in the host lattice.

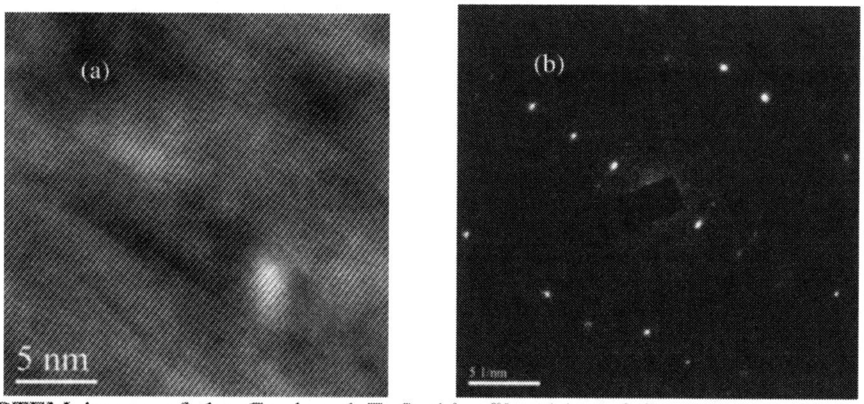

Figure 2. HRTEM image of the Co-doped ZnO thin film (a) and its corresponding selected area electron diffraction pattern (b).

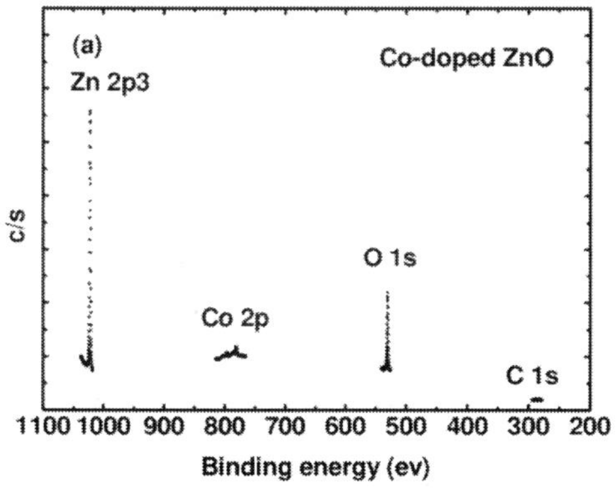

Figure 3. XPS spectra for Co-doped ZnO thin film.

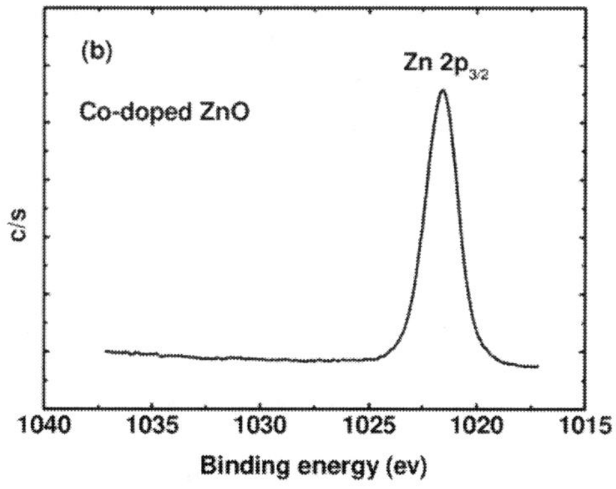

Figure 4. Zn $2p_{3/2}$ XPS spectrum for Co-doped ZnO thin film.

XPS was used to characterize the chemical states of components for the films. XPS spectra of the Co-doped ZnO film were taken in the range till 1100 eV, as shown in Figure 3. Peaks corresponding to Zn, Co and O is observed in the figure. The XPS spectrum of C 1*s* is also shown in this figure, as the binding energy of C 1*s* is usually used as an internal reference in the spectrum [15]. The spectra corresponding to O 1*s* core level, Zn 2*p* core level and Co 2*p* core level of the Co-doped ZnO specimen are calibrated by 284.6 eV binding energy of C 1*s* as the internal reference spectra. Then element atom ratio can be obtained by the spectra. Under our experimental conditions, the atom ratio of Co:Zn is about 1:10, which is much larger than that prepared in target. And the atom ratio of Zn:O is about 1:0.9, indicating that there is some oxygen vacancies in the specimen.

The Zn 2*p* spectrum is shown in Figure 4. A single Gaussian was nicely fit and the peak appeared at 1021.7 eV, which is lying within the standard reference value of Zn in ZnO. It can be identified as Zn $2p_{3/2}$ peak [16]. The binding energy value suggests that Zn atoms are in +2 oxidation state. However, based on this binding energy value, it is hard to rule out the possibility of the presence of Zn interstitial defects completely, as the differences of the binding energy mentioned above are so small [16].

The binding energy of O 1*s* is given in Figure 5. The O1*s* peak is observed to reveal a shoulder at higher binding energy. The asymmetry nature of the curve reflects the variation of oxidation number of oxygen. By fitting the peak with gaussians, the O1*s* can be separated into two peaks, the lower energy peak and the higher energy peak. The lower energy peak located at 530.2 eV (Peak 1) is in agreement with the binding energy for O 1*s* of 528.1 - 531.05 eV for a metal oxide [16], which may be attributed to oxide ions in ZnO or CoO in our case. Here, it is attributed to O – Zn(Co) bonds from the wurtzite structure of Zn^{2+} ion of the Co-doped ZnO. While the higher energy peak, located at 531.2 eV (Peak 2), corresponds to the oxygen deficient region within the ZnO matrix, that is, it can be assigned to O - H bonds resulting from exposure to air and surface contamination which is due to the chemisorbed oxygen of the surface hydroxyl, absorbed H_2O and so on. The presence of the higher energy peak coincides with the results of wide survey scan.

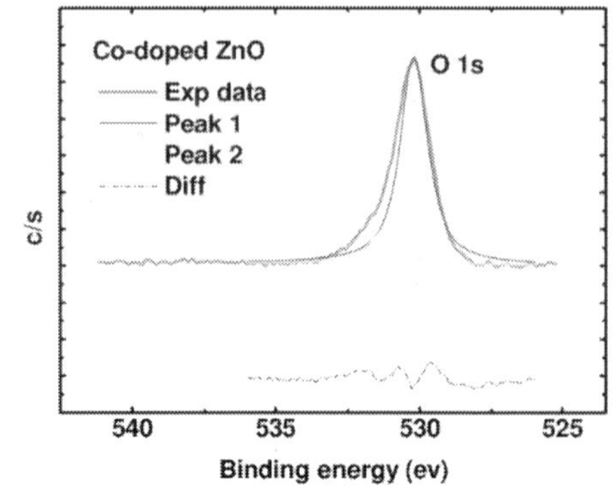

Figure 5. O 1*s* XPS spectrum for Co-doped ZnO thin film of specimen and its Curve fittings.

Figure 6 reveals the Co 2*p* XPS spectrum for Co-doped ZnO thin film in the upper panel. By fitting the peak with Gaussians (some components of Lorentzian), the spectra could be deconvoluted into four peaks P1, P2, P3 and P4. The peaks P1 located at 780.9 eV, corresponding to the binding energy of Co - O bond and can be assigned to Co 2*p*3/2 core level. It is further notable that no apparent peak occurred around 778 eV, indicating that no apparent Co - Co bonding exists in the film. Namely, the observed Co 2*p*3/2 peak position is quite different from that of the Co metal and also away from other cobaltic oxide [16]. The peak P3 around 796.4 eV, can be assigned to Co 2*p*1/2 core level. The difference between the peaks for Co 2*p*3/2 and Co 2p1/2 is close to 15.5 eV, which coincides with that of the value of CoO (around 15 eV) [16]. Hence it is reasonable to suggest that Co^{2+} incorporate into wurtzite lattice at the Zn^{2+} sites, without forming obvious impurity phase like metallic Co.

As we know, photoelectrons were ejected while XPS is processing. When the outgoing photoelectron simultaneously interacts with a valence electron and excites it or shakes it up to a higher energy level, the energy of the core electron is then reduced slightly giving a satellite structure. Satellite structure is also named as shaking up resonance transition. This perturbation gives rise to some discrete energy-loss phenomena [15] which are related to electronic structures. Hence under our experimental conditions, it is reasonable to assign the peaks P2 and P4, observed around 786.2 eV and 802.3 eV, respectively, to the satellite structures of Co 2*p*3/2 and Co 2*p*1/2, which are related to the *d*-band metal Co due to the interaction between the core hole and the valence electrons O 2*p* resulting in a change in the potential seen by the valence electrons upon the creation of the core hole. Namely, the satellite peaks reflect the excitation of a valance-band electron from an occupied state to an unoccupied state. Here, in order to analyze the photoemission spectra, the line spectra have been broadened by Gaussians and Lorentzians (mostly Gaussians). The former represents the instrumental broadening and finite bandwidth effects, and the latter relates to the lifetime broadening that increases with binding energies. The relative intensity and energy of the satellite to the main peak are carefully determined. The relationship between the relative binding energies and the ionization cross-section is schematically shown in the lower panel of Figure 6. It can be seen that strong satellite structures were obtained in this study. Furthermore, the strong and well defined satellite peaks reflect the fact that the substituted Co ions are in the 2+ oxidation states, and the interaction between cobalt and oxygen is strong, which will be discussed in details as follows.

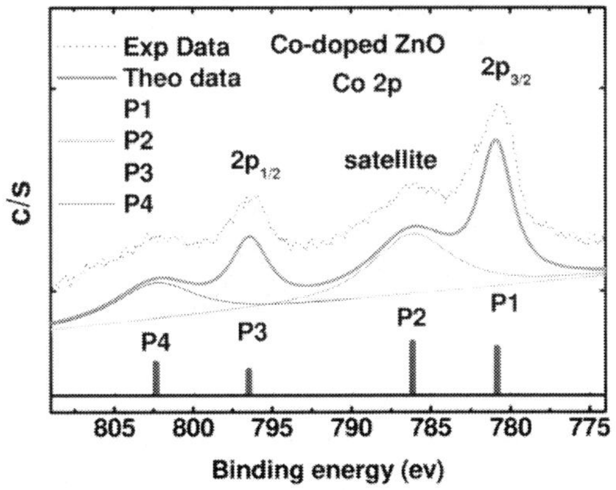

Figure 6. Co 2p core-level photoemission spectrum of Co-doped ZnO thin film and its Curve fittings shown in the upper panel. The ionization cross-section of corresponding peaks which are normalized to that of Co 2p3/2 core-level shown in the lower panel.

All these experimental results probably supply us a reasonable picture of the chemical states for the specimen. In short, the oxygen ions are lying in metal oxide, which are due to O - Zn(Co) bonds, though there are a few oxygen deficiencies and oxygen vacancies. Co ions have replaced part of Zn sites in the host lattice, resulting in Co - O bondings and no obvious precipitate was detected in the film. The substituted Co ions are in the 2+ oxidation state. More electronic-structure parameters can be induced as follows:

Our experimental results show that Co-doped ZnO thin film has a wurstzite-type lattice, that is, the thin film has a similar band structure to that of ZnO. Here, it is reasonable to consider that E_F locates in the center of the bandgap, and the relative position of Fermi level and the top of the valence band, $E_F - E_v$ for $Zn_{1-x}Co_xO$ is that $E_F - E_v \approx 1.6$ eV, similar to that of ZnO [17]. In the film, Co atoms substitute Zn-sites, hence there is a tetrahedral crystal field applied on the Co ions. The Co-doped ZnO system can be regarded as the case of Co^{2+} (d^7) in a tetrahedral field of ZnO. The exchange interaction between the *p* band in the valence band and the *d* electrons is mainly derived from the *p-d* hybridization. Especially at the Γ point, the top of the valence band is constructed purely from the anion *p* orbitals that can only hybridize with the *d* orbitals of t_2 symmetry [18].

In DMSs, two fundamental interaction mechanisms are considered: the direct Coulomb exchange and the kinetic exchange. It is known that the final state of *d* electrons is determined by the competition of crystal field splitting energy and intro-atomic exchange energy [19]. Kinetic exchange, i.e., the hybridization of the band states with localized ionic *d* orbitals, may be viewed as a second-order perturbation effect involving virtual transition of an electron between the band states and ionic orbitals [20]. The exchange interaction between the *s(p)* electrons in the conduction band and the *d* electrons of the Co^{2+} is derived from the direct exchange. To reveal the 3*d*-3*d* Coulomb interaction and the 3*d*-anion hybridization, the electronic-structure parameters Δ, U are important, where Δ, the multiplet-averaged charge-transfer energy between the ligand and the transition-metal atom, is defined as the difference between the multiplet-averaged energies of the d^n and $d^{n+1}L^1$ configurations at the transition-metal atom, where d^n denotes *n*-electron occupancy at the transition-metal site in the presence of fully filled ligand levels, and L denotes a hole in an anion p orbital. Therefore, $\Delta = E^{av}(d^{n+1}L^1) - E^{av}(d^n)$. U, the multiplet-averaged Coulomb interaction, is defined by $U \equiv E(d^{n-1}) + E(d^{n+1}) - 2E(d^n)$[21]. Based on the above expressions and the peak positions from Figure 6, relative electronic-structure parameters for Co^{2+} in Co-doped ZnO can be suggested as follows, Δ about 5.3 eV, U about 4.9 eV, Based on the devised electronic-structure parameters for Co^{2+} in Co-doped ZnO, the exchange interaction in the $Zn_{1-x}Co_xO$ thin films could be estimated, which will be discussed in our outcoming papers.

4. Conclusions

This paper addresses chemical states of the films prepared using pulsed-laser-deposition method by photoelectron spectroscopy. Our present experimental results show that it is wurtzite structure with the C-axis of the film aligned with that of the substrate without obvious precipitation. The oxygen ions are lying in metal oxide, which are due to O - Zn(Co) bonds, though there are a few oxygen deficiencies and oxygen vacancies. Co ions have replaced part of Zn sites in the host lattice, resulting in Co - O bonds. The substituted Co ions are in the 2+ oxidation state. The electronic-structure parameters Δ, U are also induced.

Acknowledgment

The authors thank Innovation Research Team for Spintronic Materials and Devices of Zhejiang Province. The work has also been supported by fund from State Key Lab of Silicon Materials, Zhejiang University (Grant No. SKL2008-5).

References

[1] Ohno H, Shen A, Matsukura F, Oiwa A, Endo A, Katsumoto S and Iye Y 1996 *Appl. Phys. Lett.* **69** 363
[2] Dietl T, Ohno H, Matsukura F, Cibert J and Ferrand D 2000 *Science* **287** 1019
[3] Wang J X, Sun X W, Yang Y, Huang H, Lee Y C, Tan O K and Vayssieres L 2006 *Nanotechnology* **17** 4995
[4] Ramón T Z, Jamil E and Claude L C 2008 *Appl. Phys. Lett.* **93** 233119
[5] Nakahara K, Takasu H, Fons P, Iwata K, Yamada A, Matsubara K, Hunger R and Niki S 2001 *J. Cryst. Growth* **227-228** 923
[6] Hsu H S, Huang J C A and Huang Y H 2006 *Appl. Phys. Lett.* **88** 242507
[7] Das J, Pradhan S K, Sahu D R, Mishra D K, Sarangi S N, Nayak B B, Verma S and Roul B K 2010 Physica B **405** 2492.
[8] Xu X Y and Cao C B 2010 *Journal of Alloys and Compounds* **501** 265
[9] Zhou S Q, Potzger K, Xu Q Y, Talut G, Lorenz M, Skorupa W, Helm M, Fassbender J, Grundmann M and Schmidt H 2009 *Vacuum* **83** S13
[10] Seghier D and Gislason H P 2009 *Physica B* **404** 4800
[11] Singhal R K, Arvind S, Xing Y T, Sudhish K, Dolia S N, Deshpanded U P, Shripathid T and Elisa B S 2010 *Journal of Alloys and Compounds* **496** 324
[12] Peng YZ, Liew T, Chong T C, An C W and Song W D 2006 *Appl. Phys. Lett.* **88** 192110

[13] Peng YZ, Liew T, Song W D and Chong T C 2009 *Chinese Physics B* **18** 5501

[14] Peng YZ, Liew T, Ye Z Z and Zhang Y Y 2007 *Chinese Science Bulletin* **52** 2742

[15] John F Watts 1990 *An introduction to surface analysis by electron spectroscopy* (Oxford University Press, Royal Microscopical Society, UK)

[16] Moulder J F, Stickle W F, Sobol P E and Bomben K D 1995 *Handbook of X-ray Photoelectron Spectroscopy* (Physical Electronics, USA)

[17] Ley L, Pollak R A, McFeely F R, Kowalczys S P and Shirley D A 1974 *Physical Review B* **9** 600

[18] Mizokawa T and Fujimori A 1997 *Physical Review B* **56** 6669

[19] Bhide V G and Rajoria D S and Mössbauer 1972 *Physical Review B* **6** 1021

[20] Blinowski J and Kacman P 1992 *Physical Review B* **46** 12298

[21] T. Mizokawa, T. Nambu, A. Fujimori, T. Fukumura and M. Kawasaki 2002 *Physical Review B* **65** 085209.

Journal of Physics: Conference Series **263** (2011) 012019 doi:10.1088/1742-6596/263/1/012019

Spin-bias-induced transport through a quantum dot coupled to a ferromagnetic lead

Yuan Li[1], **Ru Bai**[2], **Dexuan Huo**[3], **and Zhenghong Qian**[2]
[1]Department of Physics , Hangzhou Dianzi University, Hangzhou, P. R. China
[2]College of Electronic Information, Hangzhou Dianzi University, Hangzhou, P. R. China
[3] Institute of Materials Physics , Hangzhou Dianzi University, Hangzhou, P. R. China
E-mail: liyuan@hdu.edu.cn

Abstract. We investigate the spin-bias-induced transport through a quantum dot coupled to a ferromagnetic lead and a nonmagnetic semiconductor lead. Electron transport through this quantum dot is analyzed theoretically by means of the master equation method. It is shown that the polarization direction and the magnitude of the spin bias can be detected through adjusting the energy level of the quantum dot. The effects of the temperature and the tunneling rates on the charge current are also investigated numerically. The spin-bias-induced current can be easily measured in systems with low temperature and high tunneling rates. Even though the tunneling rate is small due to the conductance mismatch between the quantum dot and the ferromagnetic lead, the spin bias can still be detected through measuring the charge current.

1. Introduction

Spin based electronics or spintronics [1, 2] aims at exploiting electron spin to develop a new generation of electronic devices. The spin current is one of the essential concepts in spintronics. Thus it becomes an important task to generate, manipulate, and detect the spin current for further advancement of spintronic devices. The manipulation of the spin current, the flow of electron spins in a solid, is the key spintronics technology that will allow the achievement of efficient magnetic memories and computing devices. Many recent works have proposed some methods to generate the spin current, such as pumping excitation [3], optical injection [4, 5], magnetic tunneling injection [6] and spin-Hall effect [7]. Kato *et al.* reported the first experimental observation of the spin-current induced spin accumulation via the magneto-optical Kerr effect in GaAs semiconductor systems [8]. The spin current can also be observed through conversing it into an electric current via the inverse process of the spin-Hall effect [9]. In addition, some works proposed to indirectly measure the spin current through measuring the spin torque [10] or the spin-induced electric field [11].

A spin accumulation induced by the spin current can generate a spin-dependent splitting of chemical potentials or spin bias in nonmagnetic materials. The spin bias is regarded to be the driving force of the spin current. Since many of the intensively investigated spin-injected nonmagnetic materials were widely used to fabricate electrodes probing semiconductor [12], and single-molecular quantum dots [13]. The spin transport induced by spin bias in quantum dots were theoretically investigated [14, 15]. On the other hand, the materials and geometries including ferromagnets can not only serve as effective injectors, and also act as a part of the detector of spin bias. A small ferromagnet attached to the outlet of a spin filter may be used

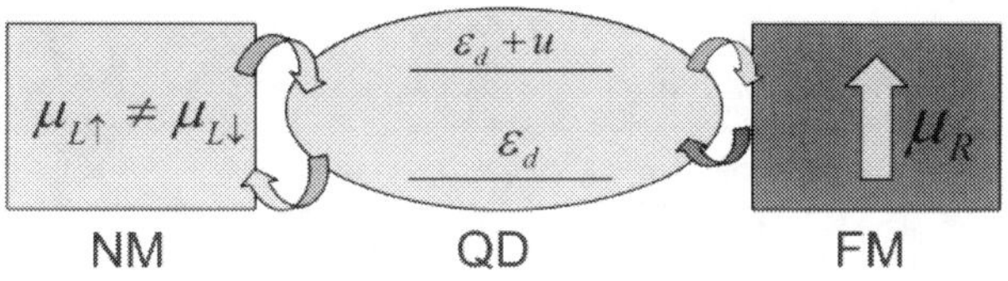

Figure 1. Schematic diagram of the device under consideration where the quantum dot is weakly coupled to one nonmagnetic semiconductor lead (NM) existing a spin bias and a ferromagnetic lead (FM).

as a detector of spin bias owing to its spin-dependent transparent coefficients if one can reduce the effect of conductance mismatch and disorder-induced scattering.

As discussed above, both the spin-bias-induced transport in nonmagnetic materials and the spin transport in the geometries including ferromagnets were already extensively studied. However the interplay of lead's ferromagnetism and spin bias in the nonmagnetic lead remains to a large extent unexplored. Therefore, in this paper we consider the spin-bias-induced transport through a quantum dot coupled to a ferromagnetic lead and a nonmagnetic lead. We use the master equation method to calculate the charge current flowing through the quantum dot. Thus the polarization direction and the amplitude of the spin bias can be detected through tuning the energy level of the quantum dot. And the effects of the temperature and the tunneling rates are also analyzed numerically in detail. It is shown that the spin-bias-induced current can be easily measured in systems with high tunneling rates in low temperature. Even though the tunneling rate is small due to the conductance mismatch between the quantum dot and the ferromagnetic lead, the spin bias can still be detected through measuring the charge current.

The paper is organized as follows. In Sec. 2, the model for the quantum dot coupled to a ferromagnetic and a nonmagnetic lead and the general formalism for the master equation method are presented. According to this method, we calculate the spin-bias-induced charge current and the electron occupation numbers of the quantum dot. In Sec. 3, We numerically analyze the occupation numbers and the charge current as a function of the spin bias, and discuss the effect of the temperature and the tunneling rates on the charge current. Final conculsions and summary appear in Sec. 4.

2. Theoretical description

The considered system consists of a single-level quantum dot coupled to left nonmagnetic semiconductor lead and right ferromagnetic lead. A spin bias in the nonmagnetic semiconductor lead can be applied by spin injection from a spin-polarized quantum point contact in combination with magnetic focusing. The general device structure we consider is schematically shown in Figure 1. The Hamiltonian can be written as $H = H_l + H_r + H_d + H_t$, with H_l and H_r describing the left nonmagnetic semiconductor lead and the right ferromagnetic lead,

$$H_\nu = \sum_{k\sigma} \varepsilon^{\nu}_{k\sigma} a^{\dagger}_{\nu k\sigma} a_{\nu k\sigma}, \tag{1}$$

for $\nu =$ L and R, where $\varepsilon^{\nu}_{k\sigma}$ is the single-electron energy in lead ν with the wave vector $\mathbf{k}$ and spin σ ($\sigma =\uparrow, \downarrow$), and $a^{\dagger}_{\nu k\sigma}$ and $a_{\nu k\sigma}$ are the corresponding creation and annihilation operators, respectively. H_d is the Hamiltonian for the interacting quantum dot, and is of the form

$$H_d = \sum_{\sigma} \varepsilon_d d^{\dagger}_{\sigma} d_{\sigma} + U d^{\dagger}_{\uparrow} d_{\uparrow} d^{\dagger}_{\downarrow} d_{\downarrow}, \tag{2}$$

where ε_d is the single energy of the quantum dot, $d_\sigma^\dagger$ (d_σ) creates (annihilates) an electron in the quantum dot, and U denotes the on-site Coulombic repulsion. The tunneling processes are described by the Hamiltonian

$$H_t = \sum_{k\sigma\nu\in \mathrm{L,R}} (t_{\nu\sigma} d_\sigma^\dagger a_{\nu k\sigma} + \mathrm{H.c.}), \tag{3}$$

where $t_{\nu\sigma}$ is the tunneling parameter, and it was assumed that electron spin is conserved in the tunneling processes.

In the sequential tunneling regime, the tunneling of an electron from source to drain electrodes is a sequence of two incoherent processes. Therefore we can use the master equation method to describe spin-bias-induced transport through the interacting quantum dot. Firstly we define four physical quantities P_0, $P_\uparrow$, $P_\downarrow$ and P_d, which describe the probabilities of the four states in the quantum dot, namely the empty state, the spin-up and spin-down occupied states, and the doubly occupied state. The occupation probabilities can be related by the normalization condition $P_0 + P_\uparrow + P_\downarrow + P_d = 1$. According to the master equation method developed by Glazman and Matweev[16], we can obtain that

$$\begin{aligned}
\frac{dP_0}{dt} &= -P_0(\Gamma_\uparrow^+ + \Gamma_\downarrow^+) + P_\uparrow\Gamma_\uparrow^- + P_\downarrow\Gamma_\downarrow^-, \\
\frac{dP_\uparrow}{dt} &= P_0\Gamma_\uparrow^+ - P_\uparrow(\Gamma_\uparrow^- + \tilde{\Gamma}_\downarrow^+) + P_d\tilde{\Gamma}_\downarrow^-, \\
\frac{dP_\downarrow}{dt} &= P_0\Gamma_\downarrow^+ - P_\downarrow(\Gamma_\downarrow^- + \tilde{\Gamma}_\uparrow^+) + P_d\tilde{\Gamma}_\uparrow^-, \\
\frac{dP_d}{dt} &= P_\uparrow\tilde{\Gamma}_\downarrow^+ + P_\downarrow\tilde{\Gamma}_\uparrow^+ - P_d(\tilde{\Gamma}_\uparrow^- + \tilde{\Gamma}_\downarrow^-),
\end{aligned} \tag{4}$$

where Γ_σ^+ (Γ_σ^-) denotes the tunneling rate of electrons with spin σ, which tunnel to (from) the single level of the quantum dot when the level is not occupied. Similarly $\tilde{\Gamma}_\sigma^+$ ($\tilde{\Gamma}_\sigma^-$) also refers to the tunneling rate of electrons associating with the tunneling processes when the single level is already occupied by an electron with spin $-\sigma$ in the quantum dot. These tunneling rates can be written as

$$\begin{aligned}
\Gamma_\sigma^\pm &= f_{l\sigma}^\pm \zeta_{l\sigma} + f_{r\sigma}^\pm \zeta_{r\sigma}, \\
\tilde{\Gamma}_\sigma^\pm &= \tilde{f}_{l\sigma}^\pm \tilde{\zeta}_{l\sigma} + \tilde{f}_{r\sigma}^\pm \tilde{\zeta}_{r\sigma},
\end{aligned} \tag{5}$$

where $f_{\nu\sigma}^+$ and $\tilde{f}_{\nu\sigma}^+$ are the Fermi distribution functions

$$\begin{aligned}
f_{\nu\sigma}^+ &= \frac{1}{\exp[(\varepsilon_d - \mu_{\nu\sigma})/k_BT] + 1}, \\
\tilde{f}_{\nu\sigma}^+ &= \frac{1}{\exp[(\varepsilon_d + U - \mu_{\nu\sigma})/k_BT] + 1},
\end{aligned} \tag{6}$$

where $\mu_{\nu\sigma}$ denotes the electrochemical potential of the lead ν, and $f_{\nu\sigma}^- = 1 - f_{\nu\sigma}^+$ and $\tilde{f}_{\nu\sigma}^- = 1 - \tilde{f}_{\nu\sigma}^+$. In Eq. (5), $\zeta_{\nu\sigma}$ and $\tilde{\zeta}_{\nu\sigma}$ are defined as follows

$$\begin{aligned}
\zeta_{\nu\sigma} &= 2\pi \sum_k |\, t_{\nu\sigma} \,|^2 \, \delta(\varepsilon_d - \varepsilon_{k\sigma}^\nu), \\
\tilde{\zeta}_{\nu\sigma} &= 2\pi \sum_k |\, t_{\nu\sigma} \,|^2 \, \delta(\varepsilon_d + U - \varepsilon_{k\sigma}^\nu).
\end{aligned} \tag{7}$$

Since in this paper we are only interested in a stationary solution, the equations satisfied by $P_i (i = 0, \uparrow, \downarrow, d)$ can be reduced to a simple form due to the stationary conditions $dP_i/dt = 0$.

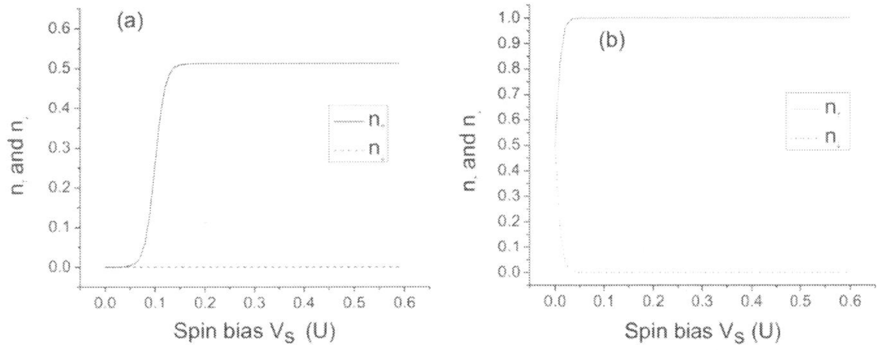

Figure 2. Occupation numbers $n_\uparrow$ (solid line) and $n_\downarrow$ (dashed line) as a function of the spin bias V_s, calculated for (a) $\epsilon_d = 0.1$ and (b) $\epsilon_d = -0.9$ in the case $\mu_{L\uparrow} = -\mu_{L\downarrow} = V_s$. Other parameters are $\zeta_L = 0.05$, $p_r = 0.9$, $\gamma = 0.5$, $\mu_R = 0$, $U = 1$ and $T = 0.01$.

After obtaining the explicit forms of the probabilities P_i through solving the stationary equations, we can calculate all the electron occupation numbers by $n_\sigma = P_\sigma + P_d$ and the charge current flowing from the quantum dot to the right ferromagnetic lead, namely,

$$I(t)/e = \sum_\sigma [P_\sigma f^-_{r\sigma}\zeta_{r\sigma} - P_0 f^+_{r\sigma}\zeta_{r\sigma} + P_d \tilde{f}^-_{r\sigma}\tilde{\zeta}_{r\sigma} - P_{-\sigma}\tilde{f}^+_{r\sigma}\tilde{\zeta}_{r\sigma}], \tag{8}$$

with e denoting the absolute value of the electron charge.

3. Numerical results

In this section we numerically analyze the spin-bias-induced transport through the quantum dot coupled to a ferromagnetic lead and a nonmagnetic semiconductor lead. The transport problems considered in this paper follow from the spin dependence of tunneling processes described by the parameters $\Gamma^\pm_\sigma$ and $\tilde{\Gamma}^\pm_\sigma$. For simplicity, we will neglect all effects following from the band structure, and assume $\zeta_{\nu\sigma} = \tilde{\zeta}_{\nu\sigma}$ in the following numerical calculation. According to the above derivation, we can analyze the transport behavior of this system in the two cases $\mu_{L\uparrow} > \mu_{L\downarrow}$ and $\mu_{L\downarrow} > \mu_{L\uparrow}$. And the right lead is ferromagnetic and the chemical potential μ_R is set to be zero. One can tune the energy level of the quantum dot ϵ_d thus to change spin current of this system. We start with calculating the occupation number of the quantum dot. In the following numerical calculations, we set the on-site Coulombic repulsion $U = 1$ as the energy unit. We also set $\hbar = e = k_B = 1$, and then the temperature T is in unit of energy. We choose $\zeta_{L\uparrow} = \zeta_{L\downarrow} = \zeta_L = 0.05$, and adopt the spin asymmetry factor p_r for the right barrier $\zeta_{R\uparrow/\downarrow} = \gamma\zeta_L(1 \pm p_r)$ to describe the spin dependence more quantitatively, where the upper (lower) sign is for the spin-majority ($\sigma =\uparrow$) and spin-minority ($\sigma =\downarrow$) electrons, and γ denotes the asymmetry between the left and right barriers. The system's parameters are chosen as $p_r = 0.9$, $\gamma = 0.5$, $\mu_R = 0$, and $T = 0.01$. According to experimental measurement [17], the on-site Coulombic repulsion is about 10mev, then $T = 0.01$ in unit of U corresponds to $T = 1K$ for $k_B T \approx 0.1$mev.

While $\epsilon_d < 0.1$, as shown in Figure 2(a), there is no electron in the quantum dot associating with $n_\uparrow = n_\downarrow = 0$. When the spin bias approaches to 0.1, the occupation number of the spin-up electron $n_\uparrow$ abruptly increases from 0 to about 0.5. While $n_\downarrow$ still remains to be zero due to $\mu_{L\downarrow} < \mu_R < \epsilon_d$. Supposed the level ϵ_d tuned to be -0.9, there exist spin-up and spin-down electrons in the quantum dot while $V_s = 0$ seen from Figure 2(b). With increment of the spin bias, $n_\uparrow$ abruptly increases to 1, while $n_\downarrow$ decreases to zero. That is to say, in this stage, the

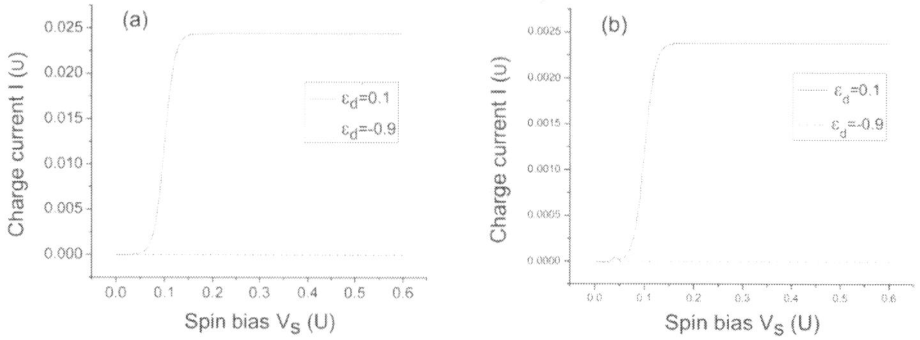

Figure 3. The charge current I as a function of the spin bias V_s, calculated for (a) $\mu_{L\uparrow} = -\mu_{L\downarrow} = V_s$ and (b) $\mu_{L\downarrow} = -\mu_{L\uparrow} = V_s$. Other parameters are the same as Figure 2.

spin-up electron can occupy the lower energy ϵ_d because ϵ_d, $\epsilon_d + U < \mu_{L\uparrow}$, then the spin-down level is pushed up to be $\epsilon_d + U > \mu_{L\downarrow}$, and so it is empty. According to numerical analysis, the variable trends of n_σ are different in the case $\mu_{L\downarrow} = -\mu_{L\uparrow} = V_s$. The variation trend of $n_\downarrow$ is similar with $n_\uparrow$'s as shown in Fiure 2, namely the spin-down electron will occupy the energy level ϵ_d and the spin-up level is empty.

Next we start with analyzing the spin-bias-induced current in the two case $\mu_{L\uparrow} = -\mu_{L\downarrow} = \pm V_s$. According to the stationary solutions of P_i, we can calculate the charge current I in terms of Eq. (8). As shown in Figure 3(a), there is no current flowing through the quantum dot while $\epsilon_d < 0.1$ for $\mu_{L\downarrow} < \mu_{L\uparrow} < \epsilon_d$. However, the current will abruptly increase to about 0.025 when the spin bias $V_s > 0.1$. The variation of the charge current can be explained in terms of the analysis of the occupation number of the quantum dot. While $V_s > 0.1$, the spin-up electron can tunnel to the quantum dot from the left lead and flow to the right ferromagnetic lead which contribute to the large current. After the level ϵ_d is tuned to be -0.9, the spin-up electron will occupy the level ϵ_d and $n_\uparrow = 1$ which can block the spin-up current. While the spin-down level is pushed up to be $\epsilon_d + U > \mu_{L\downarrow}$, hence the spin-down electron can not tunnel into the quantum dot. For the case $\mu_{L\downarrow} = -\mu_{L\uparrow} = V_s$, the charge current can only increase to about 0.0025 for $V_s > 0.1$ and $\epsilon_d = 0.1$, see Figure 3(b). In comparison to the previous case, the maximum value of the current is a lower-order quantity. Hence one can detect the polarization direction of the spin bias in terms of the amplitude of the charge current through tuning the level ϵ_d. Accordingly, the amplitude of the spin bias can also be determined in terms of the abrupt change of the charge current.

In order to investigate the transport property of this model in detail, we consider the effects of the temperature and the tunneling rates on the charge current. Figure 4 shows variation trends of the charge current as the function of the energy level ϵ_d in the two cases $\mu_{L\uparrow} = -\mu_{L\downarrow} = \pm 0.5$ for different temperatures. Let us, first, analyze the case $\mu_{L\uparrow} = -\mu_{L\downarrow} = 0.5$, see from Figure 4(a). While $\epsilon_d < 0$, the charge current approaches to zero for $T = 0.002$ and $T = 0.02$ describing by the black square and red dot curves because the level $\epsilon_d < \mu_R$.

With increment of the temperature, for example $T = 0.2$ and $T = 0.6$, there exists a small charge current induced by the effect of the temperature though the level $\epsilon_d < \mu_R$. Once the energy level $\epsilon_d > 0$, the charge current abruptly increases to about 0.025 due to $\mu_R < \epsilon_d < \mu_{L\uparrow}$. While for higher temperatures $T = 0.2$ and $T = 0.6$, the charge current I is suppressed and smaller than the case of low temperature. It is expected that the charge current can approach to zero in the regime $\epsilon_d > 0$ with enough high temperature. Hence it is suggested to measure the spin-bias-induced current in the low temperature, for example $T < 0.02$.

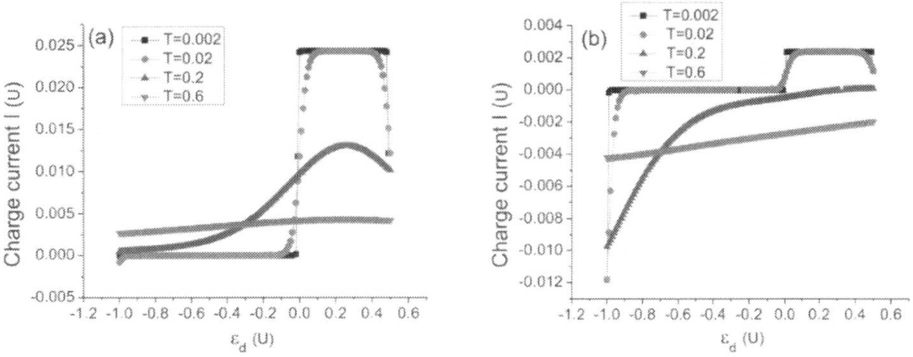

Figure 4. The current I as a function of the energy level ϵ_d for (a) $\mu_{L\uparrow} = -\mu_{L\downarrow} = 0.5$ and (b) $\mu_{L\downarrow} = -\mu_{L\uparrow} = 0.5$. The black square, red dot, blue triangle and green triangle curves are for $T = 0.002$, $T = 0.02$, $T = 0.2$ and $T = 0.6$, respectively. Other parameters are $\zeta_L = 0.05$, $p_r = 0.9$, $\mu_R = 0$, $U = 1$.

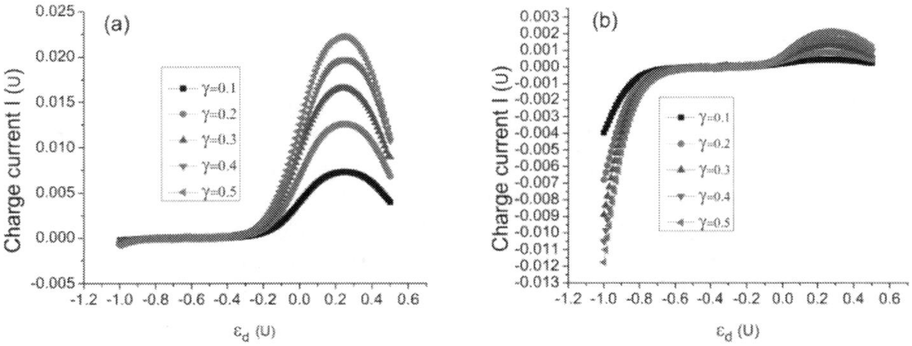

Figure 5. The current I as a function of the energy level ϵ_d for (a) $\mu_{L\uparrow} = -\mu_{L\downarrow} = 0.5$ and (b) $\mu_{L\downarrow} = -\mu_{L\uparrow} = 0.5$. The black square, red dot, blue triangle, green triangle and purple triangle curves are for $\gamma = 0.1$, $\gamma = 0.2$, $\gamma = 0.3$ $\gamma = 0.4$ and $\gamma = 0.5$, respectively. Other parameters are $\zeta_L = 0.05$, $p_r = 0.9$, $\mu_R = 0$, $U = 1$ and $T = 0.08$.

Next we turn to analyze the second case $\mu_{L\downarrow} = -\mu_{L\uparrow} = 0.5$ shown in Figure 4(b). Firstly let us discuss the energy-dependent relation of the charge current for the low temperature $T = 0.002$ and $T = 0.02$. It is obvious that there exists a large current associating with the special energy level $\epsilon_d = -1$. This phenomenon can be explained in terms of the previous analysis of the occupation number in the quantum dot. While $\epsilon_d = -1$, the spin-down electron can occupy this level because ϵ_d, $\epsilon_d + U < \mu_{L\downarrow}$, then the spin-up level is pushed up to be $\mu_R = \epsilon_d + U > \mu_{L\uparrow}$. Therefore the spin-up electron can flow from the right ferromagnetic lead to the quantum dot which contributes to the large current. If the energy level is tuned to the regime $-1 < \epsilon_d < 0$, the charge current equals to zero. When $\epsilon_d > 0$, there occurs a small charge current associating with the spin-down electron flowing through the quantum dot to the right ferromagnetic lead. The amplitude of the charge current is a lower-order quantity compared to the case $\mu_{L\uparrow} = -\mu_{L\downarrow} = 0.5$ because the spin-down electron is reflected strongly in the interface between the quantum dot and the ferromagnetic lead. Similarly the charge current is also temperature-dependent in this case.

For the device considered in this paper, the spin-bias-induced current can be affected by the tunneling rates between the quantum dot and two leads. It is necessary to investigate the effect of the tunneling rate describing by $\zeta_{\nu\sigma}$. Figure 5 shows the charge current I as a function of the energy level ϵ_d for $\gamma = 0.1 \sim 0.5$, where the coefficient γ describing the difference of the tunneling rates between $\zeta_{L\sigma}$ and $\zeta_{R\sigma}$ with $\zeta_{R\uparrow/\downarrow} = \gamma\zeta_L(1 \pm p_r)$ for two leads. We define the tunneling rate of the right lead as $\zeta_R = (\zeta_{R\uparrow} + \zeta_{R\downarrow})/2$, then the ratio between tunneling rates of the right and left leads can be calculated, i.e., $C = \zeta_R/\zeta_L = \gamma$. As illustrated in Figure 5(a), the charge current begins to increase from zero while the energy level $\epsilon_d > -0.3$ arising from the effect of the temperature with $T = 0.08$. Obviously, the charge currents are different for different tunneling rates. The maximum amplitude of the charge current is about 0.0075 for $\gamma = 0.1$. With increment of the tunneling rate, the charge current can increase to about 0.025 associating with $\gamma = 0.5$. That is to say, the spin-bias-induced current can be enhanced through increasing the tunneling rate between the quantum dot and the ferromagnetic lead. The similar result can be obtained through analyzing the variation trends of the charge current in the case $\mu_{L\downarrow} = -\mu_{L\uparrow} = 0.5$ shown in Figure 5(b). Finally we can conclude that it is easier for the spin-bias-induced current to be measured in the systems with low temperature and high tunneling rates. For those systems with low tunneling rates, however, the spin polarization direction of the spin bias can also be detected through measuring the charge current with different values of the energy level ϵ_d.

4. Summary and conclusions

In summary, we have investigated the electron transport induced by a spin bias through a quantum dot coupled to a ferromagnetic lead and a nonmagnetic semiconductor lead. The spin bias in the nonmagnetic semiconductor lead can be applied by spin injection from a spin-polarized quantum point contact in combination with magnetic focusing. The occupation number of electrons n_σ in the quantum dot presents different variation trends through adjusting the position of the energy level of the quantum dot for the two cases $\mu_{L\uparrow} > \mu_{L\downarrow}$ and $\mu_{L\downarrow} > \mu_{L\uparrow}$. The physical origin of the different occupation number arises from the interplay of the ferromagnetic lead and the Coulombic repulsion, which finally affect the charge current induced by the spin bias. Therefore, the charge current shows different variation trends for the two cases $\mu_{L\uparrow} > \mu_{L\downarrow}$ and $\mu_{L\downarrow} > \mu_{L\uparrow}$ when the energy level of the quantum dot is adjusted. The variable characteristic of the current can be utilized to detect the magnitude and polarization direction of the spin bias. It is found that the charge current is basically temperature-dependent. And the charge current will become larger with increment of the tunneling rates. According to numerical analysis, the spin-bias-induced current is suggested to be measured in systems with low temperature and high tunneling rates. Even with a low tunneling rate, however, the spin bias can still be detected through adjusting the energy levels of the quantum dot because the total variable trends keep consistent. Thus the model proposed in this paper may be a promising device to detect spin bias through measuring the charge current.

Acknowledgments

The work was supported by Innovation Research Team for Spintronic Materials and Devices of Zhejiang Province. This work was also supported by NSF-China under Grants No. 10847113 and by Foundation of HangZhou Dianzi University (Grant No. KYS075608078).

References

[1] Wolf S A, Awschalom D D, Buhrman R A, Daughton J M, Molnar S Von, Roukes M L, Chtchelkanova A Y, Tresger D M 2001 *Science* **294**, 1488.
[2] Zutic I, Fabian J and Sarma S Das 2004 *Rev. Mod. Phys.* **76**, 323.
[3] Watson S K, Potok R M, Marcus C M, and Umansky V 2003 *Phys. Rev. Lett.* **91**, 258301.

[4] Stevens M J, Smirl Arthur L, Bhat R D R, Najmaie A, Sipe J E, and Driel H M van 2003 *Phys. Rev. Lett.* **90**, 136603; Hübner J, Rühle W W, Klude M, Hommel D, Bhat R D R, Sipe J E, and Driel H M van 2003 *ibid.* **90**, 216601.
[5] Li J, Dai X, Shen S Q, and Zhang F C 2006 *Appl. Phys. Lett.* **88**, 162105; Cui X D, Shen S Q, Li J, Ji Y, Ge W K, and Zhang F C 2007 *ibid.* **90**, 242115.
[6] Valenzuela S O and Tinkham M 2006 *Nature* (London) **442**, 176; Kimura T, Otani Y, Sato T, Takahashi S, and Maekawa S 2007 *Phys. Rev. Lett.* **98**, 156601.
[7] Sih V, Myers R C, Kato Y K, Lau W H, Gossard A C, and Awschalom D D 2005 *Nature Physics* **1**, 31; Sih V, Lau W H, Myers R C, Horowitz V R, Gossard A C, and Awschalom D D 2006 *Phys. Rev. Lett.* **97**, 096605; Wunderlich J, Kaestner B, Sinova J, and Jungwirth T 2005 *Phys. Rev. Lett.* **94**, 047204.
[8] Kato Y K, Myers R C, Gossard A C, and Awschalom D D 2004 *Science* **306**, 1910.
[9] Saitoh E, Ueda M, Miyajima H, and Tatara G 2006 *Appl. Phys. Lett.* **88**, 182509.
[10] Mohanty P, Zolfagharkhani G, Kettemann S, and Fulde P 2004 *Phys. Rev. B* **70**, 195301; Chen T W, Chen C M, Huang C M, and Guo G Y 2006 *ibid.* **73**, 235309.
[11] Meier F and Loss D 2003 *Phys. Rev. Lett.* **90**, 167204; Schütz F, Kollar M, and Kopietz P, *ibid,* **91**, 017205; Sun Q F, Guo H, and Wang J 2004 *Phys. Rev. B* **69**, 054409; Sun Q F, and Xie X C 2005 *ibid,* **72**, 245305.
[12] 1997 *Mesoscopic Electron Transport, NATO Advanced Study Institutes, Seriers E: Applied Science Vol. 345*, ed L L Sohn and L P Kouwenhoven et al, (Kluwer, Dordrecht).
[13] Park H, Park J, Lim A K L, Anderson E H, Alivisators A P, and McEuen P L 2000 *Nature* (London), **407**, 57.
[14] Sun Q F, Xing Y X, and Shen S Q 2008 *Phys. Rev. B* **77**, 195313.
[15] Lu H Z, Shen S Q 2008 *Phys. Rev. B* **77**, 235309.
[16] Glazman L I and Matveev K A 1988 *Pis'ma Zh Éksp. Teor. Fiz.* **48**, 403 [*JETP Lett.* 1988 **48**, 445]
[17] Wiel W G van der, Franceschi S De, Elzerman J M, Fujisawa T, Tarucha S, and Kouwenhoven L P 2002 *Rev. Mod. Phys.* **75**, 1.

Journal of Physics: Conference Series **263** (2011) 012020 doi:10.1088/1742-6596/263/1/012020

Suppressed charge ordering and cluster glass behavior of $Pr_{0.75}Na_{0.25}Mn_{0.95}Fe_{0.05}O_3$

Q Cheng [1], Y Li [2] and X Luo[2]

[1] Material Science and Engineering College, Northeast Forestry University, Harbin 150040, China

[2] College of Science, Northeast Forestry University, Harbin 150040, China

E-mail: liyu87043@nefu.edu.cn

Abstract. The cusps of charge ordering and the long range AFM order of the "pseudo"-CE type are suppressed with increasing Fe concentration in $Pr_{0.75}Na_{0.25}Mn_{1-x}Fe_xO_3(0\leq x\leq 0.30)$. Ac susceptibility at low temperatures of $Pr_{0.75}Na_{0.25}Mn_{0.95}Fe_{0.05}O_3$ was investigated. The real component (χ') shows that freezing temperature T_f is linear in the logarithm of the frequency. The normalized slope $P=\Delta T_f/T_f\Delta log_{10}\omega$, which is much lower than canonical insulating spin glass systems in which $0.06\leq P\leq 0.08$. The χ' peak at T_f are suppressed with the increasing frequency. The intensity of imaginary component (χ'') at T_f and spin blocking temperature $T_B\sim 10$ K shows a complicated variation trend with increasing frequency. The ground state of the sample is cluster glass with a coexistence of charge-ordered phase, correlated ferromagnetic clusters in spin glass matrix.

1. Introduction

Mixed valence manganates $Ln_{1-x}A_xMnO_3$ (where Ln is a lanthanide and A is an alkaline earth ion or alkaline ion) have been widely researched because of their variety of properties [1-4]. Those compounds of them exhibiting charge order (CO) phenomena become the focus of intense studies because the percolative paths of coexisting ferromagnetic (FM) metallic phase are suspected to be responsible for the colossal magnetoresistance (CMR) [5]. The low temperature ordering of Mn^{3+} and Mn^{4+} can be destroyed by the cation disorder resulting from the substitution, either at Mn site [6] or at Ln-A site [7]. Introducing impurity into Mn site results in insulator-to-metal as well as antiferromagnetic-to-ferromagnetic transitions and induces the electronic phase separation of CO antiferromagnetic domain and charge-disordered FM domain [8], spin-glass and cluster glass.

The ac susceptibility for some single sample has been widely researched in order to decide the ground state of the sample [9]. The frequency shift in T_f offers a possible criterion for distinguishing a canonical spin glass from a spin-glass-like material, and from a superparamagnet. On the other hand, the variation trend of imaginary component (χ'') intensity of the ac susceptibility is a possible criterion for distinguishing a canonical spin glass from a phase separation material [10]. The ac susceptibility for $Pr_{0.75}Na_{0.25}MnO_3$ displays a cluster glass state [9]. $Pr_{0.75}Na_{0.25}Mn_{0.90}Fe_{0.10}O_3$ displays a Phase separation ground state [10]. In order to comprehend the frequency shift in T_f and the variation trend of imaginary

Journal of Physics: Conference Series **263** (2011) 012020 doi:10.1088/1742-6596/263/1/012020

component (χ'') peak with increasing frequency, Investigate ac susceptibility of different Fe doppant in $Pr_{0.75}Na_{0.25}MnO_3$ is important. In this paper, ac susceptibility at low temperatures of $Pr_{0.75}Na_{0.25}Mn_{0.95}Fe_{0.05}O_3$ was investigated. The normalized slope $P=\Delta T_f/T_f\Delta log_{10}\omega$ and the variation trend of imaginary component (χ'') of the ac susceptibility with increasing frequency for $Pr_{0.75}Na_{0.25}Mn_{0.95}Fe_{0.05}O_3$ were investigated. Some complicated variation trends were observed. The possible reason was presented. The ground state of the sample was presented.

2. Experimental

The single-phase polycrystalline perovskites $Pr_{0.75}Na_{0.25}Mn_{1-x}Fe_xO_3$ ($0 \leq x \leq 0.30$) were prepared by a sol-gel technique (See Ref 11). The room-temperature powder X-ray diffraction patterns show a single-phase orthorhombic structure for all the samples. A physical properties measurement system (PPMS, Quantum Design Inc.) was used for all the magnetic measurements.

3. Results and discussion

The temperature dependence of the zero-field cooled (ZFC) and field cooled (FC) magnetization at 0.01 T for $Pr_{0.75}Na_{0.25}Mn_{1-x}Fe_xO_3$ is shown in Figure.1. The parent phase

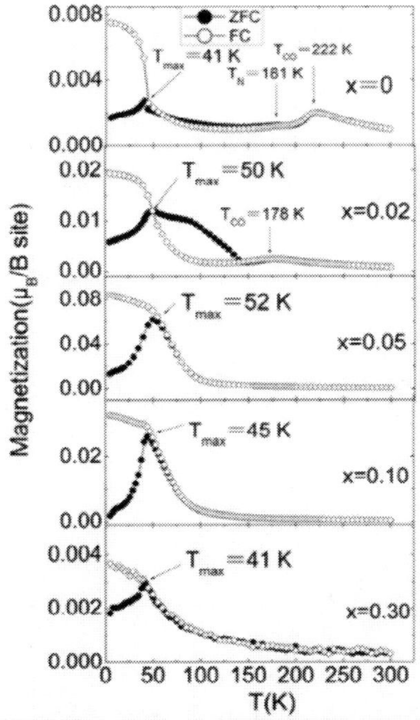

Figure. 1 Field cooled (open symbols) and zero field cooled (closed symbols) magnetizations of $Pr_{0.75}Na_{0.25}Mn_{1-x}Fe_xO_3$ ($0 \leq x \leq 0.30$) as a function of temperature, measured at 0.01 T.

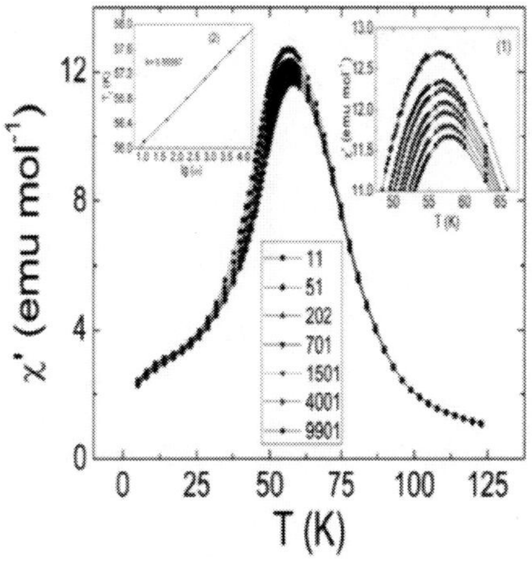

Figure. 2 The temperature dependence of real component (χ') of the ac susceptibility for $Pr_{0.75}Na_{0.25}Mn_{0.95}Fe_{0.05}O_3$ measured in a field of 0.0005 T with different frequencies. The inset of (1) shows the imaginary of the peaks, (2) shows the variation of the χ' peak temperature with the frequency of the driving ac field in the range 11-9997 Hz.

$Pr_{0.75}Na_{0.25}MnO_3$ displays a behavior similar to that reported by Takuya Satoh [12] and J. Hejtmánek [13]. The broader anomaly at ~222 K and the small "hump" at ~181 K are associated with charge ordering (CO) and a long-range antiferromagnetic order of the "pseudo"-CE type, respectively. The branches of ZFC and FC bifurcate below ~ 41 K. The cusps of CO and the long range AFM order of the "pseudo"-CE type are suppressed with increasing Fe concentration. This sample series (except for the x = 0.02 sample) exhibit similar behavior to the parent phase sample at low temperatures. The x = 0.02 sample shows a more complicated behavior. The ZFC magnetization increases sharply below 140 K and cannot superpose on FC magnetization. The ZFC magnetization nearly levels off between 90 K and 50 K. Then the magnetization begins to decrease as the temperature is further lowered, which is similar to the ZFC magnetization of $Pr_{0.5}Ca_{0.5}Mn_{0.975}Al_{0.025}O_3$ [13].

The temperature dependence of real component of ac susceptibility of $Pr_{0.75}Na_{0.25}Mn_{0.95}Fe_{0.05}O_3$ for different frequencies is shown in Figure. 2. The intensity of the peak in $\chi'(T)$ decreases and shifts upwards in temperature with increasing frequency, which is also qualitatively consistent with either a spin glass [14-16] or a cluster system [17-20]. But it can be quantified through the frequency dependence of the freezing temperature (T_f) as given by the peak position in $\chi'(T)$. As shown in the inset (2) of Figure. 2, T_f is linear in the logarithm of the frequency with a normalized slope $P=\Delta T_f/T_f\Delta log_{10}\omega$. Normalized slope $P=$ 0.01 for this sample is much lower than canonical insulating spin glass systems in which $0.06\leq P\leq 0.08$. ($Pr_{0.75}Na_{0.25}Mn_{0.95}Fe_{0.05}O_3$ has an insulating ground state below ~65 K [11]). Although the P value for spin glass systems is $0.0045 \leq P \leq 0.08$ [14]. This frequency shift in T_f offers a possible criterion for distinguishing a canonical spin glass from a spin-glass-like material, and from a superparamagnet.

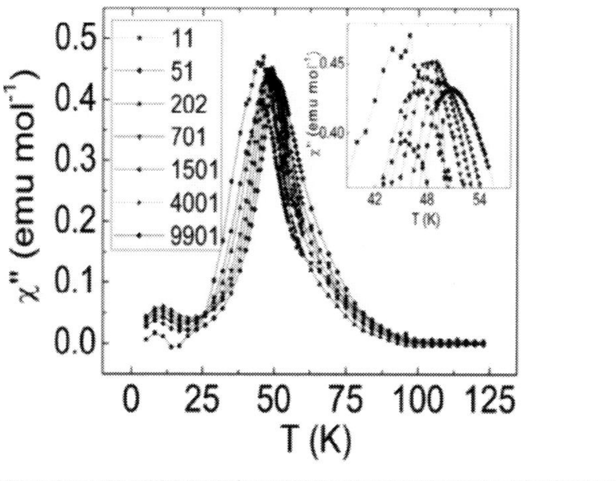

Figure. 3 The temperature dependence of Imaginary component (χ'') of the ac susceptibility for $Pr_{0.75}Na_{0.25}Mn_{0.95}Fe_{0.05}O_3$ measured in a field of 0.0005 T with different frequencies. The inset shows the magnify of the peak.

The temperature dependence of Imaginary component (χ'') of the ac susceptibility for $Pr_{0.75}Na_{0.25}Mn_{0.95}Fe_{0.05}O_3$ measured in a field of 0.0005 T with different frequencies are shown in Figure. 3. The inset shows the magnify of the peak. χ'' T_f shows an even more complicated behavior. The peak value decrease with increasing frequency for f ≤ 52 Hz and increase subsequently till 1501 Hz, and then decrease with further increasing frequency. The results of χ'' give a stronger evidence that the low field ground state for the sample series does not behave as a conventional spin glass, nor a phase separation. Perhaps, the ground state of the sample is cluster glass with a coexistence of charge-ordered phase, correlated ferromagnetic clusters in spin glass matrix. The x = 0.05 ratio of the components for Fe doping induces the complicated ac susceptibility variation trend. The other factor affect the ac susceptibility perhaps is the difference of the size of the ferromagnetic clusters in the x = 0.05 sample from the x = 0 and x = 0.10 sample. The larger distribution in size of the ferromagnetic clusters induces the more complicated ac susceptibility variation trend. The peak at T_{max} of ZFC M-T curve measured at 0.01 T field for the x = 0.05 sample is more broad that suggests the magnetic clusters has a larger distribution in size in the field, and its ac susceptibility variation trend is more complicated.

A small maximum at ~10 K in χ'' (T) is found as shown in Figure. 3. It is corresponding to the inflexion in M (T) and χ' (T) at the same temperature as shown in Figure. 1 and Figure. 2, respectively. It could be related to the blocking of isolated spins between ferromagnetic clusters as has been suggested for similar features in traditional spin glass materials [21] and phase-separated material [9]. The maximum at ~10 K for the x=0.05 sample increase with increasing frequency when f ≤ 4001 Hz and then decrease with further increasing frequency. This is also different to the x = 0, 0.10 sample, which with a variation trend of the maximum at ~10 K with increase frequency reverse to the variation trend of the maximum at T_f.

4. Conclusions

The cusps of CO and the long range AFM order of the "pseudo"-CE type are suppressed with increasing Fe concentration in $Pr_{0.75}Na_{0.25}Mn_{1-x}Fe_xO_3(0 \leq x \leq 0.30)$. Ac susceptibility at low temperatures of $Pr_{0.75}Na_{0.25}Mn_{0.95}Fe_{0.05}O_3$ was investigated. The intensity of real component (χ') at T_f are suppressed with the increasing frequency and shows that freezing temperature T_f is linear in the logarithm of the frequency with a normalized slope $P=\Delta T_f/T_f\Delta log_{10}\omega$, which is much lower than canonical insulating spin glass systems in which $0.06 \leq P \leq 0.08$. The intensity of imaginary component (χ'') at T_f decrease with increasing frequency for f ≤ 52 Hz and increase subsequently till 1501 Hz, and then decrease with further increasing frequency. The maximum at ~10 K in χ'' (T) increase with increasing frequency when f ≤ 4001 Hz and then decrease with further increasing frequency. The ground state of the sample is cluster glass with a coexistence of charge-ordered phase, correlated ferromagnetic clusters in spin glass matrix.

Acknowledgements
This work was supported by the Open Project of Key Laboratory for Magnetism and Magnetic Materials of the Ministry of Education, Lanzhou University (Grant No.LZUMMM2010004) the Fundamental Research Funds for the Central Universities (Grant No.DL09BB31) and the Fundamental Research Funds for the Central Universities (Grant No.DL09BB36).

References

[1] P. Schiffer, A. P. Ramirez, W. Bao, S.-W. Cheong 1995 Phys. Rev. Lett. **75** 3336.
[2] A. P. Ramirez, P. Schiffer, S-W. Cheong, C. H. Chen, W. Bao, T. T. M. Palstra, P. L.

Gammel, D. J. Bishop, and B. Zegarski 1996 Phys. Rev. Lett. **76** 3188.
[3] C.H. Chen and S-W. Cheong 1996 Phys. Rev. Lett. **76** 4042.
[4] J.hejtmánek, Z.Jirák, J.sebek, A. Śtrejc, M. Hervieu 2001 J. Appl. Phys. **89** 7413.
[5] M. Uehara, S. Mori, C.H. Chen, S. -W. Cheong 1999 Nature (London) **399** 560.
[6] F. Damay, A. Maignan, C. Martin, and B.Raveau 1997 J. Appl. Phys. **82** 1485.
[7] F. Damay, C. Martin, A. Maignan, B. Raveau 1997 J. Appl. Phys. **82**(12) 6181.
[8] J. Xu, Y. Matsui, T. Kimura, Y. Tokura 2001 physica C. **357-360** 401.
[9] I. G. Deac, J. F. Mitchell, P. Schiffer 2001 Phys. Rev. B **63** 172408.
[10] Yu Li, Jipeng Miao, Yu Sui, Zhe Lü, Zhengnan Qian, Wenhui Su 2006 J. Magn. Magn. Mater. **305** 247.
[11] Yu Li, Jipeng Miao,Yu Sui, Xianjie Wang, Wei Zhang, Yuqiang Liu, Ruibin Zhu, and Wenhui Su 2008 J. Alloys Comp **458** 11.
[12] J.hejtmánek, Z.Jirák, J.sebek, A. Śtrejc, M. Hervieu 2001 J. Appl. Phys. 89 7413.
[13] Takuya Satoh, Yshinao Kikuchi, Kenjiro Miyano, Emil Pollert, Jiri Hejtmánek, Zdenek Jirák 2002 Phys. Rev. B **65** 125103.
[14] J. A. Mydosh 1993 *Spin Glasses: An Experimental Introduction* (Taylor& Francis, London), p. 42.
[15] S. Chikuzami 1997 *Physics of Ferromagnetism* (Clarendon, Oxford,), p. 293.
[16] G. Williams 1991 in *Magnetic Susceptibility of Superconductors and Other Spin Systems*, edited by R. A. Hein, T. L. Francavilla, D. H. Liebenberg (Plenum, New York,) p. 475, and references therein.
[17] J. L. Tholence 1991 in *Magnetic Susceptibility of Superconductors and Other Spin Systems*, edited by R. A. Hein, T. L. Francavilla, D. H. Liebenberg (Plenum, New York,), p. 503.
[18] J.L. Dormann, R. Cherkaoui, L. Spinu, M. Noguès, F. Lucari, F.D'Orazio, D. Fiorani, A. Garcia, E. Tronc, J.P. Jolivet 1998 J. Magn. Magn. Mater. **187** 139.
[19] H. Mamiya, I. Nakatani, T. Furubayashi 1998 Phys. Rev. Lett. **80** 177.
[20] J. A. De Toro, M. A. López de la Torre, J. M. Riveiro, R. Sáez Puche, A. Gómez-Herrero, L. C. Otero-Díaz 1999 Phys. Rev. B **60** 12 918.
[21] L.M. Fisher, A.V. Kalinov, I.F. Voloshin, N.A. Babushkina, K.I. Kugel, D.I. Khomskii 2003 J. Magn. Magn. Mater. **258-259** 306.

Journal of Physics: Conference Series **263** (2011) 012021 doi:10.1088/1742-6596/263/1/012021

Co_2FeAl films with perpendicular magnetic anisotropy in multilayer structure

X Q Li, X G Xu*, S Q Yin, D L Zhang, J Miao, and Y Jiang*

State Key Laboratory for Advanced Metals and Materials, School of Materials Science and Engineering, University of Science and Technology Beijing, Beijing 100083, China.

E-mail: xgxu@ustb.edu.cn; yjiang@ustb.edu.cn.

Abstract. We have fabricated Co_2FeAl (CFA) films with perpendicular magnetic anisotropy (PMA) in a $(Co_2FeAl/Ni)_6$ multilayer structure. The effects of underlayer Cu thickness (t_{Cu}), Co_2FeAl thickness (t_{CFA}) and Ni thickness (t_{Ni}) on the magnetic properties have been studied. The PMA is realized with a large anisotropy energy density $K=3.7\times10^6$ ergs/cm^3, a high squareness $M_r/M_s=1$ and a small perpendicular coercivity $H_c=60$ Oe, while t_{Cu}, t_{CFA} and t_{Ni} are 9 nm, 0.2 nm and 0.6 nm respectively. The PMA remains after 300 ℃ annealing, which demonstrates better thermal stability of the $(Co_2FeAl/Ni)_6$ multilayer than that of $(Co/Ni)_n$.

1. Introduction

Ferromagnets with perpendicular magnetic anisotropy (PMA) are promising candidates for applications in ultrahigh recording density devices, because the key problems of superparamagnetic effect and vortex magnetization could be solved and lower critical currents for switching as well as high thermal stability can be achieved by using PMA materials [1], which include rare-earth/transition-metal alloys [2]-[4], $L1_0$-ordered (Fe, Co)–Pt [5], [6] alloys and Co/(Ni, Pt, Pd) multilayers [1], [7]. However, there is a big drawback for most PMA materials that their spin polarizations are low, causing small tunneling magnetoresistance (TMR) or giant magnetoresistance (GMR).

Half metallic ferromagnets (HMFS) have a 100% theoretical spin polarization due to their unusual band structures [8]. Therefore, it could be important if one can realize perpendicularly magnetized HMFS. Recently, the Co_2FeAl (CFA) films with PMA were obtained by co-sputtering with Tb [9]. In this method, the CFA exists in $L2_1$ structure, which is believed to have both a high spin polarization and high Curie temperature [8], [10]. The Co_2MnSi (CMS) film with PMA has been reported by Hiratsuka et al. [11]. The CMS film was introduced as an inserted layer to enhance the spin polarization of magnetic tunnel junctions (MTJs). However, in all these works, the PMA in CFA or CMS films was induced by exchange coupling with other PMA materials. In this study, we observe the PMA in $(CFA/Ni)_6$ multilayer. The PMA shows better thermal stability than that of $(Co/Ni)_n$ multilayer.

2. Experiment

The films were deposited on Si/SiO_2 substrates using conventional dc sputtering. The base pressure in the chamber was better than 1.0×10^{-5} Pa and the sputtering pressure was 0.5 Pa. The stacking of the structure is $Si/SiO_2/Ta(2)/Cu(t_{Cu})/[CFA(t_{CFA})/Ni(t_{Ni})]_6/Ta(2)$, (layer thickness in nanometer, t_{Cu}, t_{CFA} and t_{Ni} represent the thickness of Cu underlayer, CFA and Ni layers, respectively). Magnetic hysteresis loops were measured by an alternating gradient magnetometer (AGM) at room temperature.

3. Results and Discussion

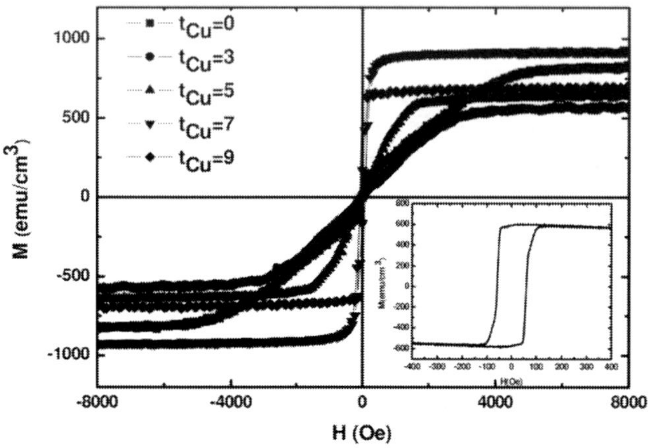

Figure 1. Out of plane M-H loops for the $Si/SiO_2/Ta(2)/Cu(t_{Cu})/[CFA(0.2)/Ni\ (0.6)]_6/Ta\ (2)$ structures with different t_{Cu}. The inset shows the low-field loop for the sample with t_{Cu}=9 nm.

Figure 1 shows the hysteresis loops of the $Si/SiO_2/Ta(2)/Cu(t_{Cu})/[CFA(0.2)/Ni(0.6)]_6/Ta(2)$ films with different t_{Cu}. While t_{Cu} increases from 0 to 7 nm, the perpendicular saturation field (H_s) decreases from 5200 Oe to 1200 Oe, which demonstrates the transformation from in-plane anisotropy to PMA. When t_{Cu} increases to 9 nm, a typical PMA is observed with a high remanence ratio of nearly 1 and a low perpendicular coercivity field (H_c) about 60 Oe. The similar phenomenon that a Cu underlayer with proper thickness can efficiently promote PMA has ever been reported in Co/Ni multilayers [12].

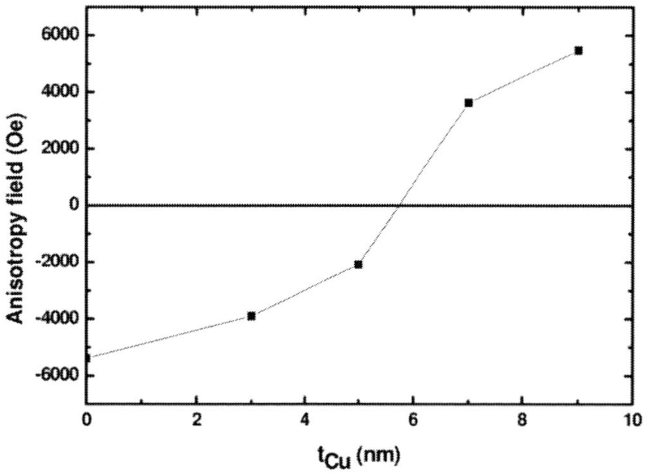

Figure 2. Anisotropy fields as a function of t_{Cu} for the $Si/SiO_2/Ta(2)/Cu(t_{Cu})/[CFA(0.2)/Ni(0.6)]_6/Ta(2)$ multilayers. Positive anisotropy fields correspond to out-of-plane magnetization.

Figure 2 shows the dependence of anisotropy field (H_{an}) on t_{Cu}. H_{an} refers to the saturation field measured under in-plane (out of plane) applied field for PMA (in-plane anisotropy) materials. The H_{an} increases monotonically with the increasing t_{Cu} and the largest H_{an} is obtained at t_{Cu}= 9 nm. The

Journal of Physics: Conference Series **263** (2011) 012021 doi:10.1088/1742-6596/263/1/012021

anisotropy energy density K can be calculated using the relationship $K=M_sH_{an}/2$, where M_s refers to the saturation magnetization. When t_{Cu}=9 nm, K is 3.7×10^6 ergs/cm^3, which is two times larger than the sample with t_{Cu}= 7 nm (K=1.8×10^6 ergs/cm^3). It is noteworthy that the values of M_s (680 emu/cm^3) and K (3.7×10^6 ergs/cm^3) for the sample with t_{Cu}= 9 nm are very similar to those observed in (Co/Ni) multilayers (M_s =650 emu/cm^3, K=3.3×10^6 ergs/cm^3) [7]. It indicates that the mechanism of PMA is similar between the (CFA/Ni) and (Co/Ni) multilayers.

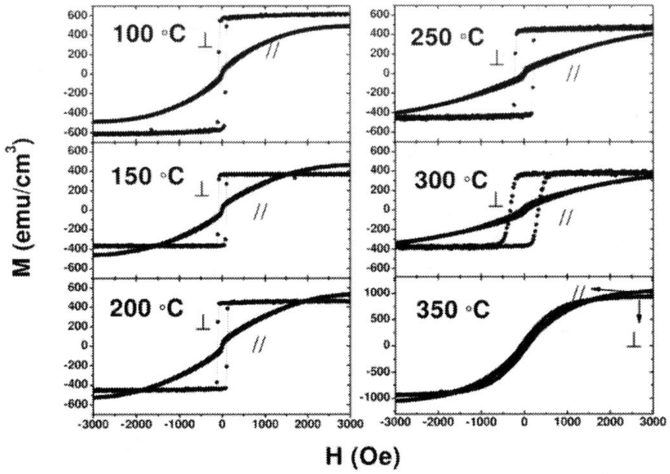

Figure 3. M-H loops under various annealing temperatures for the Si/SiO$_2$/Ta(2)/Cu(9)/[CFA(0.2)/Ni(0.6)]$_6$/Ta(2) multilayer.

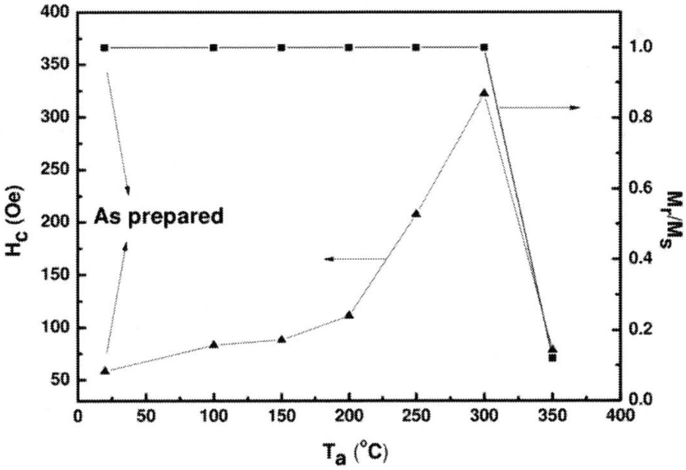

Figure 4. H_c and squareness (M_r/M_s) as a function of Ta for the Si/SiO$_2$/Ta(2)/Cu(9)/[CFA(0.2)/Ni(0.6)]$_6$/Ta(2) multilayer.

In order to study the thermal stability of the (CFA/Ni) multilayer, the heat treatment was carried on for the sample Si/SiO$_2$/Ta(2)/Cu(9)/[CFA(0.2)/Ni(0.6)]$_6$/Ta(2) under different annealing temperature (T_a) in a vacuum furnace below 2.0×10^{-4} Pa for 0.5 h. The results are shown in figure 3. The sample can maintain PMA after annealing at 300 ℃. While T_a is 350 ℃, both H_s measured in out of plane and in-plane directions are very large, which means the PMA disappears after annealing. There is a prominent feature observed from figure 3 that the PMA is enhanced significantly as T_a increases from 100 ℃ to 300 ℃. The H_c and squareness (M_r/M_s) as a function of T_a are plotted in figure 4. When T_a increases from room temperature to 300 ℃, the H_c strongly depends on T_a and the M_r/M_s remains constant (~1). After annealing at 350 ℃, both H_c and M_r/M_s drop down dramatically due to the disappeared PMA. The interface mixing under high temperature annealing should be responsible for the deterioration of PMA. However, the thermal stability of the CFA/Ni multilayers is better than

Co/Ni multilayers in which the PMA cannot maintain after 300 ℃ annealing [13]. The stronger PMA in CFA/Ni multilayers is because more symmetry breaking surface/interface anisotropy which favors PMA may be induced by different lattice constants.

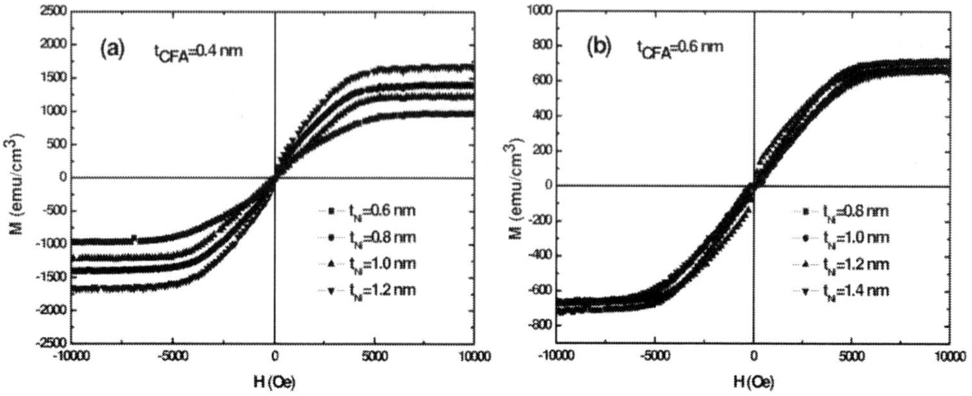

Figure 5. Out of plane M-H loops for the $Si/SiO_2/Ta(2)/Cu(9)/[CFA(t_{CFA})/Ni(t_{Ni})]_6/Ta(2)$ multilayers with different t_{Ni}. (a) t_{CFA}=0.4 nm and (b) t_{CFA}=0.6 nm.

In addition, fixing the t_{Cu} at 9 nm, two series of $SiO_2/Ta(2)/Cu(9)/[CFA(0.4)/Ni(t_{Ni})]_6/Ta(2)$ and $SiO_2/Ta(2)/Cu(9)/[CFA(0.6)/Ni(t_{Ni})]_6/Ta(2)$ (layer thickness in nanometer) multilayers were deposited. Here, t_{Ni} ranges from 0.6 nm to 1.2 nm for the former structure and the t_{Ni} ranges from 0.8 nm to 1.4 nm for the later one. No apparent PMA can be observed for all the samples with t_{CFA}=0.4 and 0.6, as shown in figure 5. It suggests that the PMA is very sensitive to the thickness of both CFA and Ni layers. Because the origin of PMA is induced by surface/interface effects, the maximum CFA thickness is limited to a value small enough. It is similar to the (Co/Ni) or (Co/Pt) multilayers with PMA in which the maximum Co thickness is usually limited to less than 1 nm [14].

4. Conclusions

In summary, we have successfully achieved perpendicular magnetization in the structure of $Si/SiO_2/Ta(2)/Cu(9)/[CFA(0.2)/Ni(0.6)]_6/Ta(2)$. The as-prepared multilayer with the structure shows the typical PMA with a large anisotropy energy density $K=3.7\times10^6$ ergs/cm^3, a high squareness $M_r/M_s=1$ and a small perpendicular H_c=60 Oe. The PMA maintains after 300 ℃ annealing, which demonstrates better thermal stability than that of $(Co/Ni)_n$ multilayers. The PMA is very sensitive to the thickness of both CFA and Ni layers.

Acknowledgments

This work was partly supported by the NSFC (Grant No. 50701005, 50831002), the Keygrant Project of Chinese Ministry of Education (No. 309006), the National Basic Research Program of China (Grant No. 2007CB936202), and the Fundamental Research Funds for the Central Universities.

References

[1] Tadisina Z R, Natarajarathinam A, Clark B D, Highsmith A L, Mewes T, Gupta S, Chen E, and Wang S 2010 J. Appl. Phys. 107, 09C703.
[2] Chen Tzu-Jung, Canizo-Cabrera. A, Chang Che-Hao, Liao Kuo-An, Li Simon C, Hou Chun-Kan, and Wu Te-ho 2006 J. Magn. Magn. Mater. 304, e291.
[3] Liu Xiaodong, Xu Zhen, Gao Ruixin, Hu Haining, Chen Zhifeng, Wang Zixin, Du Jun, Zhou Shiming, and Lai Tianshu 2008 Appl.phys.Lett. 92, 232501.

[4] Lee Ching-Ming, Ye Lin-xiu, Hsieh Tung-Hsien, Huang Chao-Yuan, and Wu Te-Ho 2010 J. Appl. Phys. 107, 09C712.
[5] Yoshikawa M, Kitagawa E, Nagase T, Daibou T, Nagamine M, Nishiyama K, Kishi T, and Yoda H 2008 IEEE Trans. Magn. 44, 2573.
[6] Kim G, Sakuraba Y, Oogane M, Ando Y, and Miyazaki T 2008 Appl. Phys. Lett. 92, 172502.
[7] Mangin S, Ravelosona D, Katine J A, Carey M J, Terris B D, and Fullerton E E 2006 Nature Mater. 5, 210.
[8] Okamura S, Miyazaki A, Sugimoto S, Tezuka N, and Inomata K 2005 Appl. Phys. Lett. 86, 232503.
[9] Li X Q, Xu X G, Zhang D L, Miao J, Zhan Q, Jalil M B A, Yu G H, and Jiang Y 2010 Appl. Phys. Lett. 96, 142505.
[10] Karthik S V, Rajanikanth A, Takahashi Y K, Okhubo T, and Hono K 2006 Appl. Phys. Lett. 89, 052505.
[11] Hiratsuka T, Kim G, Sakuraba Y, Kubota T, Kodama K, Inami N, Naganuma H, Oogane M, Nakamura T, Takanashi K, and Ando Y 2010 J. Appl. Phys. 107, 09C714.
[12] Li Zhenya, Zhang Zongzhi, Zhao Hui, Ma Bin, and Jin Q Y 2009 J. Appl. Phys. 106, 013907.
[13] Dahmane Y, Auffret S, Ebels U, Rodmacq B, and Dieny B 2008 IEEE Trans. Magn. 44, 2865.
[14] Nistor L E, Rodmacq B, Auffret S, and Dieny B 2009 Appl. Phys. Lett. 94, 012512.

Ferromagnetism studies of Cu-doped and (Cu, Al) co-doped ZnO thin films

S Z Wu, H L Yang, X G Xu[1], J Miao, Y Jiang[1]

State Key Laboratory for Advanced Metals and Materials, School of Materials Science and Engineering, University of Science and Technology Beijing, Beijing 100083, China.

Email: xgxu@ustb.edu.cn; yjiang@ustb.edu.cn.

Abstract. We have studied the room temperature ferromagnetism (FM) in Cu-doped and (Cu, Al) co-doped ZnO thin films which were grown on quartz substrates by chemical method based on a sol-gel process combining with spin-coating technology. X-ray diffraction (XRD) patterns demonstrate that both the Cu-doped and (Cu, Al) co-doped ZnO films have the hexagonal wurtzite structure with c-axis orientation. Alternating Gradient Magnetometer (AGM) measurements confirm that all the doped ZnO samples are ferromagnetic at room temperature. When the doped Cu content is 1 %, the Cu-doped ZnO film has the strongest FM. The FM significantly decreases in the (Cu, Al) co-doped ZnO films. The doping of Al ions suppresses the FM induced by the doped Cu ions.

1. Introduction

Integrating magnetic properties into semiconductor materials is a pivotal mission for practical application. [1] Recently diluted magnetic semiconductors (DMSs) have attracted much attention because of their potential applications in spintronic devices, which exploit the spin in magnetic materials along with the charge of electrons in semiconductors. Dietl et al [2] predicted that wide band gap materials such as ZnO should be a promising candidate for room temperature DMS with hole doping. Sato and Katayama-Yoshida [3] reported theoretically that not only the hole doped ZnO but also the other transition metal (TM) doped ZnO show room temperature ferromagnetism (FM). After that, TM-doped ZnO such as Co-doped ZnO, Fe-doped ZnO and Mn-doped ZnO [4-6] with FM above room temperature have been reported. But the origin of FM in TM-doped ZnO is controversial because the FM is possibly caused by a secondary phase or external pollution. [7, 8] In order to avoid the controversies, non-magnetic elements doped ZnO have been studied. [9-11] Cu is also an ideal doping element because it is a potential magnetic ion with a total spin of 1/2 by Hund's rule and the Cu-related secondary phases such as CuO and Cu_2O are non-ferromagnetic. [12-16] Therefore Cu-doped ZnO has the potential of exhibiting FM only due to the substitution of Cu on Zn sites. Meanwhile, many co-doped DMS materials have been studied. For instance, Zn doping can affect the concentration of Fe and optical properties in (Zn, Fe) co-doped CdS films. [17] Duan et al. [18] reported that (Co, Mn) co-doped Zn thin films synthesized by an autocombustion method show room

[1] To whom any correspondence should be addressed.

temperature FM. These results demonstrate that the co-doping may lead to remarkable change in the properties of DMS.

In this letter, we have studied the magnetic properties of Cu-doped and (Cu, Al) co-doped ZnO thin films which were prepared by sol-gel process combining with spin-coating technology. The study of the (Cu, Al) co-doped ZnO is helpful to figure out the origin of FM in Cu-doped ZnO samples.

2. Experimental details

In our experiment, we used zinc acetate 2-hydrate $Zn(CH_3COO)_2{\cdot}2H_2O$, copper acetate 1-hydrate $Cu(CH_3COO)_2{\cdot}H_2O$, aluminum nitrate 9-hydrate $Al(NO_3)_3{\cdot}9H_2O$, lactic acid and ethanol to synthesize the coating solution. All these reagents are of analytical grade and were directly used without special treatment.

The coating solution was prepared by a simple sol-gel process. $Zn(CH_3COO)_2{\cdot}2H_2O$, $Cu(CH_3COO)_2{\cdot}H_2O$ and/or $Al(NO_3)_3{\cdot}9H_2O$ were used as the raw materials and dissolved in the ethanol. The lactic acid was used as the stabilizer. The total metal ion concentration was 0.4 mol/L. Having been sufficient mixed, the coating solution formed crystal and symmetrical sol.

The coating solution was then dropped on quartz substructure and spin coated with 2000 r/m for 20s in air. This process made a precursor film on the substrate. Then the precursor film was dried at 150 °C in air for 10 min. This spin-coating process was repeated 20 times on each substrate. Then, the precursor films were annealed in air at 600 °C for 1 h with a calefactive speed of 1.8 °C /min.

We have synthesized Cu-doped ZnO thin films ($Zn_{1-x}Cu_xO$, 1% $\leq$Cu$\leq$ 7%) and (Cu, Al) codoped ZnO thin films ($Zn_{0.98}Cu_{0.01}Al_{0.01}O$, $Zn_{0.97}Cu_{0.01}Al_{0.02}O$, $Zn_{0.96}Cu_{0.02}Al_{0.02}O$ and $Zn_{0.97}Cu_{0.02}Al_{0.01}O$) by this method. In order to make a comparison, we also prepared pure ZnO and Al-doped ZnO thin films. The magnetic hysteresis loops were characterized by Alternating Gradient Magnetometer (AGM). The crystal structure was characterized by X-ray diffraction (XRD) using Cu k α radiation.

3. Results and discussion

XRD patterns of the $Zn_{1-x}Cu_xO$ and pure ZnO films are shown in figure 1, the XRD patterns of the $Zn_{1-x-y}Cu_xAl_yO$ films are shown in figure 2. XRD patterns show that the wide peaks at the positions which are high agree with the PDF pattern of ZnO (JCPDS 36-1451). The XRD patterns demonstrate that nearly all the Cu-doped, Al-doped and (Cu, Al) co-doped ZnO films exhibit hexagonal wurtzite structure while no other phase exists. All of these XRD patterns show that the peaks at 34.4° are much stronger than others, indicating that all these films have a strong c-axis orientation.

Compare with the pure ZnO samples, all the (002) peaks in the doped ZnO are slightly right shifted which illustrates the incorporation of Cu or Al ions into ZnO lattices. As we know, the radius of Cu^+, Cu^{2+}, Zn^{2+} ions in tetrahedral coordination are 0.60, 0.57, and 0.60 Å, respectively. [19] For the Cu-doped ZnO, because the Cu^+ and Zn^{2+} ions have the same radius, the doping of Cu^+ should not cause much change of the lattice parameters. In other words, the doping of Cu^+ ions will not affect the XRD pattern of ZnO. On the other hand, the substitution of Cu^{2+} for Zn^{2+} ions should reduce the lattice parameter of ZnO and make the ZnO peaks right shift. According to the XRD results, the (002) peak in the doped ZnO has a slight right shift which indicates the existence of Cu^{2+} ions. However, since the doping of Cu^+ ions will not affect the XRD pattern of ZnO, we can not demonstrate whether there exist Cu^+ ions.

Figure 3 shows the room temperature M-H loops for the pure and Cu-doped ZnO films. The pure ZnO film is superparamagnetic while all the Cu-doped ZnO films are ferromagnetic. The saturation magnetization (M_S) of the Cu-doped ZnO films decreases with the increasing content of Cu ions. The results are similar to the ones reported by several other groups. [12-15] The FM is induced by the p-d hybridization between 3d of Cu^{2+} and ZnO valence bands (O-p bands), which forms a magnetic moment. On the other hand, the increased antiferromagnetic interaction in Cu-Cu ions reduces the FM in the higher Cu doping films.

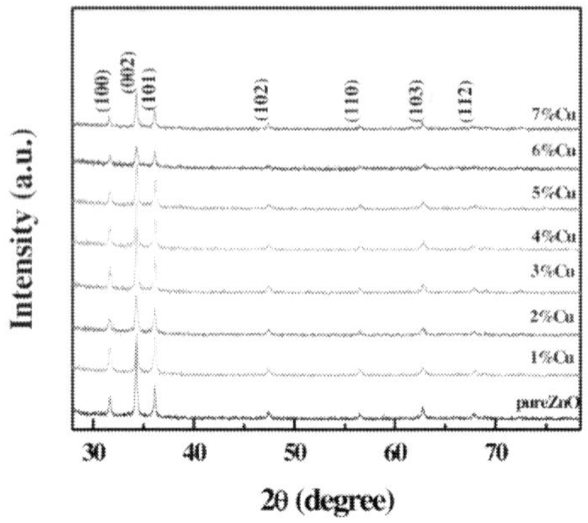

Figure 1. X-ray diffraction patterns of Cu-doped ZnO and pure ZnO films.

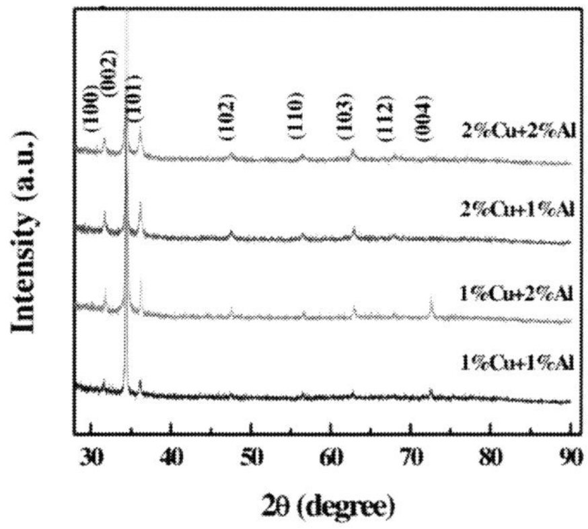

Figure 2. X-ray diffraction patterns of (Cu, Al) codoped ZnO films.

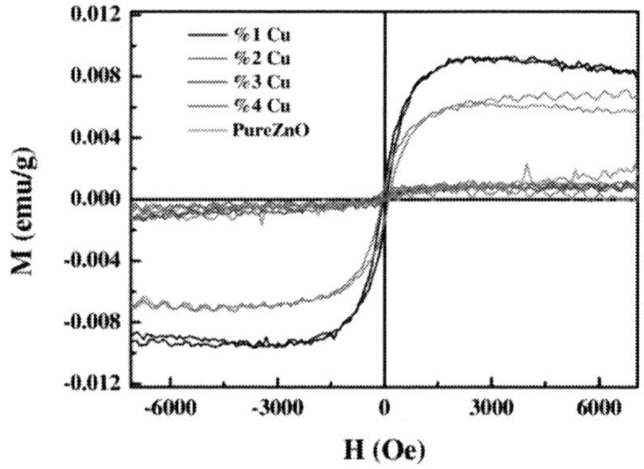

Figure 3. Room-temperature M-H loops of Cu-doped ZnO and pure ZnO films

To further study the origin of FM, we measured the hysteresis loops for the (Cu, Al) co-doped ZnO films, as shown in figure 4. All the (Cu, Al) co-doped ZnO samples are ferromagnetic. Figure 5 gives the variation of M_S with the doped Cu/Al ratio. Comparing with the Cu-doped ZnO, the FM is significantly weakened in $Zn_{0.98}Cu_{0.01}Al_{0.01}O$, $Zn_{0.97}Cu_{0.01}Al_{0.02}O$ and $Zn_{0.96}Cu_{0.02}Al_{0.02}O$ films which have Al ions not less than Cu ions. The M_S of $Zn_{0.97}Cu_{0.01}Al_{0.02}O$ film is slightly lower than that of $Zn_{0.98}Cu_{0.01}Al_{0.01}O$ film. Our result demonstrates that in the ratio of nearly 1:1 the doping of Al ions significantly suppresses the FM originating from the doping of Cu ions. Moreover, the M_S of $Zn_{0.97}Cu_{0.02}Al_{0.01}O$ film is nearly the same as that of $Zn_{0.99}Cu_{0.01}O$. Although several groups [20, 21] observed room temperature FM in Al-doped ZnO, we have not observed FM in our Al-doped ZnO. We therefore infer that Al doping decreases FM by suppressing the effect of Cu ions. One Al^{3+} ion can only affect one Cu ion because one Al^{3+} offers one free electron in ZnO lattices. According to the carrier-mediated mechanism theory, the free electrons of Al^{3+} in ZnO lattices should suppress the p-d hybridization between 3d of Cu^{2+} and ZnO valence bands (which is the origin of FM) because these free electrons may capture the 3d electrons of Cu^{2+}.

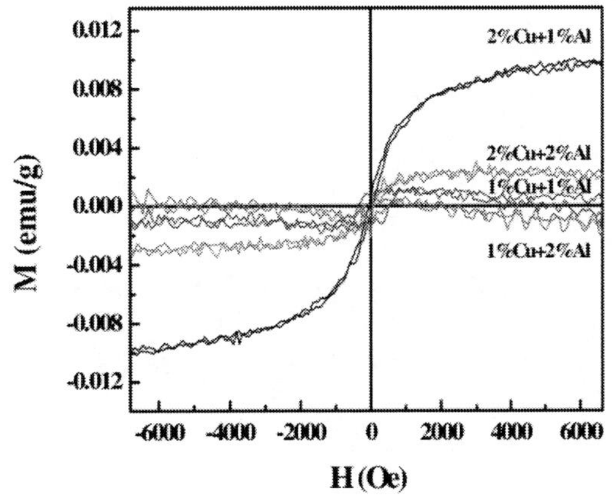

Figure 4. Room-temperature M-H loops of (Cu, Al) codoped ZnO films.

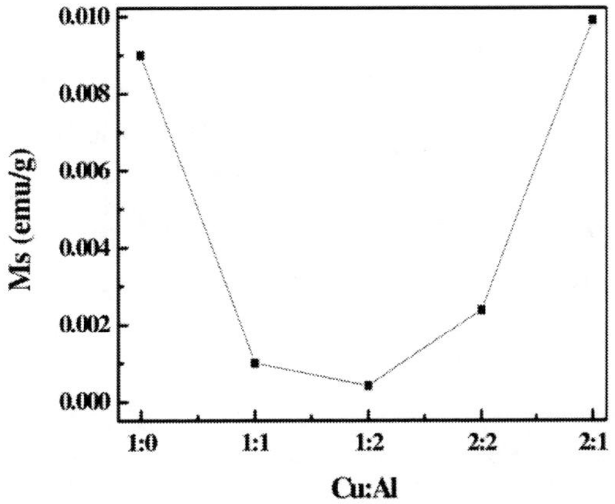

Figure 5. Room-temperature M_S of (Cu, Al) codoped ZnO films.

4. Conclusions

The Cu-doped and (Cu, Al) co-doped ZnO thin films which were grown on quartz substrates have been studied. We have observed room temperature FM in the $Zn_{1-x}Cu_xO$ and $Zn_{1-x-y}Cu_xAl_yO$ films. The experimental data show that the doped Cu ions are the primary reason for the FM. The FM has been significantly weakened in the (Cu, Al) co-doped ZnO films. The doped Al ions suppress the FM induced by the doped Cu ions, which can be understood by the carrier-mediated mechanism theory.

Acknowledgments
This work was partially supported by the National Natural Science Foundation of China (Grant Nos. 50701005, 50831002, 50971025), the National Basic Research Program of China (Grant No. 2007CB936202) and the Key Grant Project of Chinese Ministry of Education (No. 309006).

References

[1] Gary A Prinz 1998 *Science* **282** 1660
[2] T dietl, H Ohino, F Matsukura, J Cibert and D Ferrand 2000 *Science* **287** 1019
[3] K Sato and H Katayama-Yoshida 2000 *Jpn. J. Appl. Phys.* Part 2 **39** L555
[4] K Ueda, H Tabata and T Kawai 2001 *Appl. Phys. Lett.* **79** 988
[5] S D Yoon, Y Chen, D Heiman, A Yang, N Sun, C Vittoria, and V G. Harris 2006 *J. Appl. Phys.* **99** 08M109
[6] P Sharma, A Gupita, K V Rao, F J Owens, R Sharma, R Ahuja, J M Osorio Guillen, B Johansoon and A A Gehring 2003 *Nat. Mater.* **2** 673
[7] C Sudakar, J S Thakur, G Lawes, R Naik, and V M Naik 2007 *Phys. Rev. B.* **75** 054423
[8] Y Belghazi, G Schmerber, S Colis, J L Rehspringer, and A Dinia 2006 *Appl. Phys. Lett.* **89** 122504
[9] H Pan, J B Yi, L Shen, R Q Wu, J H Yang, J Y Lin, Y P Feng, J Ding, L H Van, and J H Yin 2007 *Phys. Rev. Lett.* **99** 127201
[10] Santa Chawla, K Jayanthi, and R K Kotnala 2009 *Phys. Rev. B.* **79** 125204
[11] X G Xu, H L Yang, Y Wu, D L Zhang, S Z Wu, J Miao and Y Jiang *arXiv* 1003.4423.
[12] T S Herng, S P Lau, S F Yu, H Y Yang, X H Ji, J S Chen, N Yasui and H Inaba 2006 *J. Appl. Phys.* **99** 086101
[13] D Chakraborti, J Narayan and J T Prater 2007 *Appl. Phys. Lett.* **90** 062504
[14] Xiao Li Li, Xiao Hong Xu, Zhi Yong Quan, Jun Feng Guo. Hai Shun Wu, and G A Gehring 2009 *J. Appl. Phys.* **105** 103914
[15] Prashant K Sharma, Ranu K Dutta and Avinash C Pandey 2009 *J. Magn. Magn. Mater.* **321** 4001-4005
[16] M Wei, N Braddon, D Zhi, P A Midgley, S K Chen, M G Blamire, and J L MacManus Driscoll 2005 *Appl. Phys. Lett.* **86** 072514
[17] K Liu, J Y Zhang, X Wu, B Li, Y Lu, X Fan, and D Shen 2007 *Physica B (Amsterdam)* **389** 248
[18] L B Duan, G H Rao, Y C Wang, J Yu and T Wang 2008 *J. Appl. Phys.* **104** 013909
[19] R D Shannon 1976 *Acta Crystallogr., Sect. A: Cryst. Phys., Diffr., Theor. Gen. Crystallogr.* **32** 751
[20] S J Chen, K Suzuki, and J S Garitaonandia 2009 *Appl. Phys. Lett.* **95** 172507
[21] Y W Ma, J Ding, J B Yi, H T Zhang and C M Ng 2009 *J. Appl. Phys.* **105** 07C503

Journal of Physics: Conference Series **263** (2011) 012023 doi:10.1088/1742-6596/263/1/012023

Study on the dual-synthetic antiferromagnetic property using the Co_2FeAl Heulser electrodes

D L Zhang, X G Xu *, Y Wu, X Q Li, J Miao and Y Jiang *

State Key Laboratory for Advanced Metals and Materials, School of Materials Science and Engineering, University of Science and Technology Beijing, Beijing 100083, China

E-mail: yjiang@ustb.edu.cn; xgxu@ustb.edu.cn.

Abstract. In this paper, we present the experimental results of dual-synthetic antiferromagnets (DSyAFs) with Co_2FeAl (CFA) Heusler electrodes. It is shown that when the thicknesses of Ru layers are (0.45, 0.65) and (0.45, 1.00) (in nm), the CFA-based DSyAFs have a strong synthetic antiferromagnetic coupling among three CFA layers at room temperature, with a large saturation magnetic field Hs of ~11000 Oe, a low saturation magnetization Ms of ~708 emu/cm^3 and a switching field Hsw of ~2.0 Oe, respectively. It is exciting that the CFA-based DSyAFs have an excellent thermal stability up to 400 ℃. Therefore, the CFA-based DSyAFs are favourable for applications in future spintronic devices.

1. Introduction

Spintronic devices based on spin valves (SPVs) or magnetic tunnel junctions (MTJs) such as magnetic random access memory (MRAM) and magnetic logic circuit have been considered as promising candidates for the next generation ultrahigh density magnetic storage devices [1-2]. As chief components in the SPVs or MTJs, synthetic antiferromagnets have attracted considerable attention and have been widely researched. Half-metallic ferromagnets (HMFs) have a band gap in the minority spin in band structure and theoretically exhibit a 100% spin polarization at the Fermi level [3]. Especially, the Co-based full-Heusler alloys with a chemical formula of Co_2YZ (Y = Fe, Mn; Z = Al, Si, Ge, Ga, Sn) show excellent half-metallic properties and have been considered as promising materials in spintronics devices [4-5]. Magnetostatic interaction between pinned and free layers in SPVs becomes an important problem and attracts great attentions. Synthetic antiferromagnet (SyAF) provides a method to resolve the problem due to its closed magnetic flux configuration and low stray field [6-7].

In recent years, researchers have made continuous efforts to find new synthetic antiferromagnets or optimize SyAF in order to achieve excellent antiferromagnetic coupling properties of large saturation field (H_s), low saturation magnetization (M_s), small coercivity (H_c), and good thermal stability. Due to the high spin polarization and Curie temperatures, Co-based full-Heusler alloys are ideal ferromagnetic materials used in SPVs or MTJs and have attracted much interest [8]. If the Heulser alloys can be applied in SyAF structure, it will be possible to make "all Heusler" SPVs or MTJs and get much higher giant magnetoresistance (GMR) or tunnel magnetoresistance (TMR). Xu *et al* [9] have firstly reported a CFA-based SyAF structure which shows a strong antiferromagnetic coupling between two CFA layers at room temperature (RT). However, the antiferromagnetic coupling can only maintain up to 150 ℃, indicating poor thermal stability.

Therefore, we investigate the CFA-based dual-synthetic antiferromagnet (DSyAF) structure in order to obtain excellent synthetic antiferromagnetic properties and improve thermal stability.

2. Experiment

In this paper, the CFA-based DSyAFs with a stack of Ta(4)/CFA(5)/Ru(x)/CFA(3)/Ru(y)/CFA(5)/Ta(4) (x=0.45, y=0.45; x=0.45, y=0.65; x=0.45, y=1.00 x=0.65, y=0.65; x=0.65, y=1.00; x=1.00, y=1.00) (in nm) were deposited on a Si (100) substrate using dc magnetron sputtering with an argon pressure of ~0.5 Pa. The base pressure of chamber was better than 2.0×10^{-5} Pa. A 100 nm CFA monolayer and CFA-DSyAF samples were fabricated at RT. In order to study the crystalline structure and thermal stability, all the samples were annealed at 100 ℃, 200 ℃, 300 ℃, and 400 ℃ for 1 hour in a vacuum furnace with a pressure lower than 5×10^{-5} Pa, respectively. Magnetic properties of the samples were measured by alternating gradient magnetometer (AGM), and the crystalline structure of CFA monolayer was determined by X-ray diffraction (XRD) at RT.

3. Results and discussion

Figure 1(a) shows the magnetization curves of the CFA-based DSyAFs. The results show that the CFA-based DSyAFs (x=0.45, y=0.45, 0.65 and 1.00) have a strong antiferromagnetic coupling among three CFA layers. When the thicknesses of the two Ru layers are x=0.45 nm and y=0.45 nm, the saturation field H_s, the saturation magnetization M_s and the switching field H_c are ~6000 Oe, 572 emu/cm^3, and 3.6 Oe, respectively. When x=0.45 and y increases from 0.45 nm to 1.00 nm, H_s increases from ~6000 Oe to ~11000 Oe, with an increasing M_s from 572 emu/cm^3 to 708 emu/cm^3 [as shown figure1(a)] and a decreasing H_c from 3.6 Oe to 2.0 Oe [as shown figure1(b)]. So it is obvious that the CFA-based DSyAFs (x=0.45 y=0.65 and 1.00) have an excellent synthetic antiferromagnetic property.

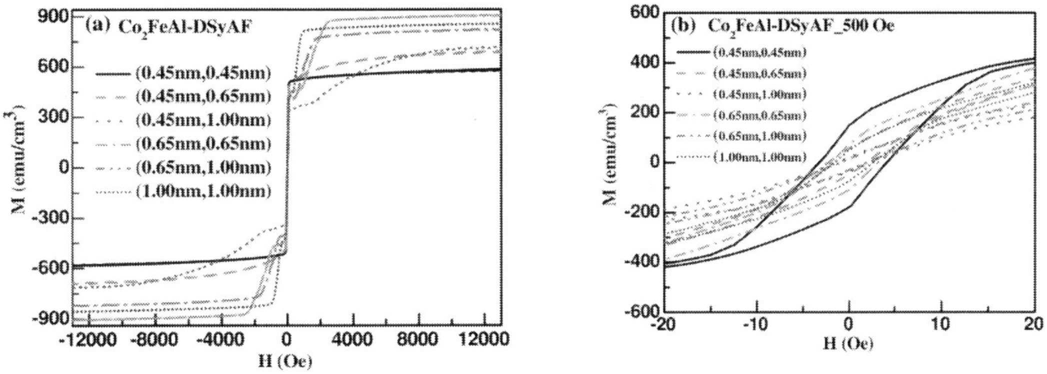

Figure 1 (a). M-H loops for as-deposited CFA-DSyAFs. (b). M-H loops for the minor loops.

Because the middle CFA layer is thinner than the others, it allows the antiferromagnetic coupling arising from the two Ru layers to be superposed on one another to enhance the overall coupling strength, thus gaining the largest saturation field H_s about ~11000 Oe compared with the H_s value of the CFA-based SyAF of only ~5000 Oe. The CFA-based DSyAF (x=0.45, y=1.00) has a saturation magnetization of ~708 emu/cm^3, which is lower than those of the $Co_{90}Fe_{10}$-based DSyAF (~870 emu/cm^3) and SyAF (~1200 emu/cm^3) [10]. The reason is that the M_s of CFA (~1000 emu/cm^3) [9] is smaller than that of $Co_{90}Fe_{10}$ (~1300 emu/cm^3). In addition, the switching field H_{sw} about ~2.0 Oe of the CFA-based DSyAF (x=0.45, y=1.00) is lower than that of the CFA-based SyAF (~4.3 Oe), especially much lower than those of the $Co_{90}Fe_{10}$-based DSyAF (~150 Oe) and SyAF (~70 Oe). This is because that the H_c of CFA (~5.0 Oe) is much smaller than that of $Co_{90}Fe_{10}$ (~57 Oe) [10].

In order to investigate the thermal stability of the CFA-based DSyAFs (x=0.45, y=0.65; x=0.45, y=1.00), the samples were annealed at 100 ℃, 200 ℃, 300 ℃, and 400 ℃ in a vacuum furnace. The M-H loops of the as-deposited and annealed CFA-based DSyAFs (x=0.45, y=0.65; x=0.45, y=1.00) are presented in Figure 3 (a~b). The results show that the H_s and M_s of CFA-based DSyAF (x=0.45, y=0.65) are about ~4492 Oe and ~510 emu/cm^3 with the annealing temperature 400 ℃ as shown in figure 2(a). It indicates that the CFA-based DSyAF has an excellent thermal stability up to 400 ℃, despite the antiferromagnetic coupling strength is weakened by the increasing annealing temperature. In addition, the H_s and M_s of CFA-based DSyAF (x=0.45, y=1.00) are about ~2510 Oe and ~701 emu/cm^3 with annealing temperature increasing to 400 ℃. It also indicates that the CFA-based DSyAF (x=0.45, y=1.00) has the good thermal stability up to 400 ℃ as shown figure 2(b).

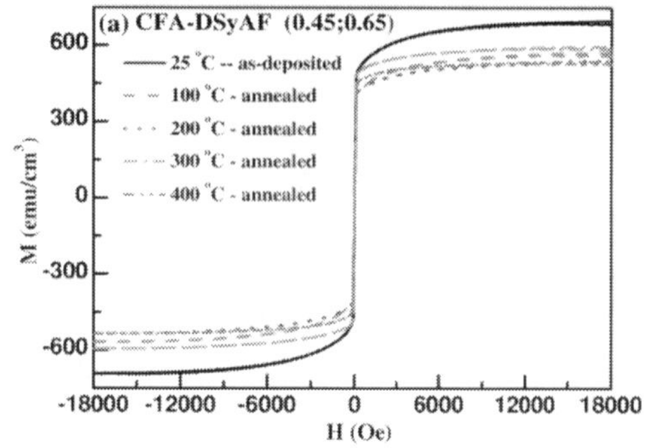

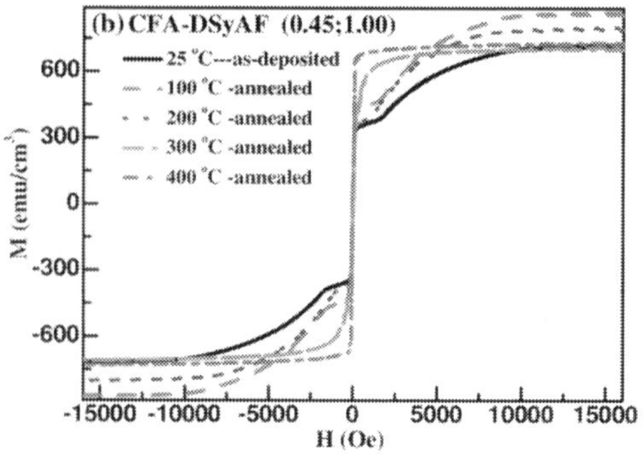

Figure 2 (a). M-H loops of CFA-DSyAF (0.45; 0.65). (b). M-H loops of CFA-DSyAF (0.45; 1.00).

We studied the crystalline structure of CFA monolayer at different annealing temperatures. Figure 3 shows the XRD patterns for the annealed CFA films. As for the curves of 300 ℃ annealed CFA films, only (220) peak of the full-Heusler alloy can be detected, indicating an A2-disordered structure. When the annealing temperature increases to 400 ℃, there emerges (200) peak of the full-Heusler alloy in the XRD pattern, which indicates that the B2-ordered structure can be obtained after annealing at 400 ℃. So the CFA-based DSyAFs with the high spin polarized B2-ordered structure perform excellent synthetic antiferromagnetic properties and thermal stability.

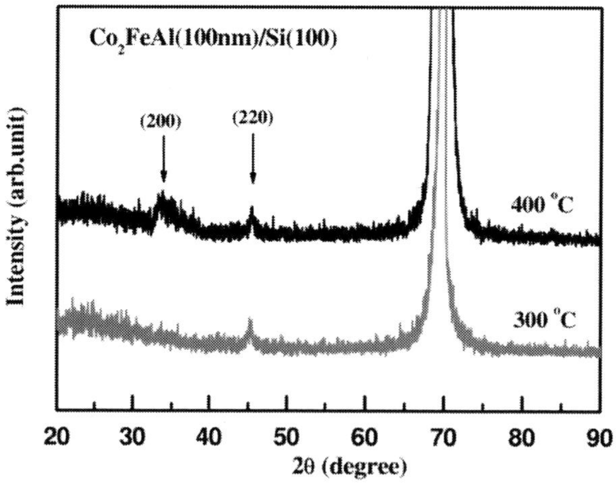

Figure 3. XRD patterns of 100 nm CFA film.

In addition, we have also studied the correlation between the H_s (M_s and H_c) and the annealing temperatures. As shown in Figure 4(a), the H_s decreases from ~13237 Oe to ~4492 Oe in the CFA-based DSyAF(x=0.45, y=0.65) and from ~12548 Oe to ~2510 Oe in the CFA-based DSyAF(x=0.45, y=1.00) with temperature increasing from 25℃ to 400 ℃. The saturation magnetization M_s increases gradually from ~708 emu/cm^3 to ~858 emu/cm^3 at T_a=100 ℃, then decreases slowly to ~701 emu/cm^3 at T_a=400 ℃ in the CFA-based DSyAF (x=0.45, y=1.00), and the M_s of the CFA-based DSyAF (x=0.45, y=0.65) decreases slowly from ~680 emu/cm^3 to ~510 emu/cm^3 with temperature increasing from 25 ℃ to 400 ℃ as shown in figure 4 (b). For the coercivity H_c, the CFA-based DSyAF (x=0.45, y=0.65 and 1.00) have a small H_c about 2.0 Oe at RT. With the temperature increasing from 25 ℃ to 400 ℃, the H_c of the CFA-based DSyAF (x=0.45, y=0.65) increases slowly from ~2.0 Oe to ~14.7 Oe. However, the H_c of the CFA-based DSyAF(x=0.45, y=1.00) increases gradually from ~2.0 Oe to 6.8 Oe under 300 ℃, and then decreases to ~1.5 Oe at T_a=400 ℃ as shown in figure 4 (c).

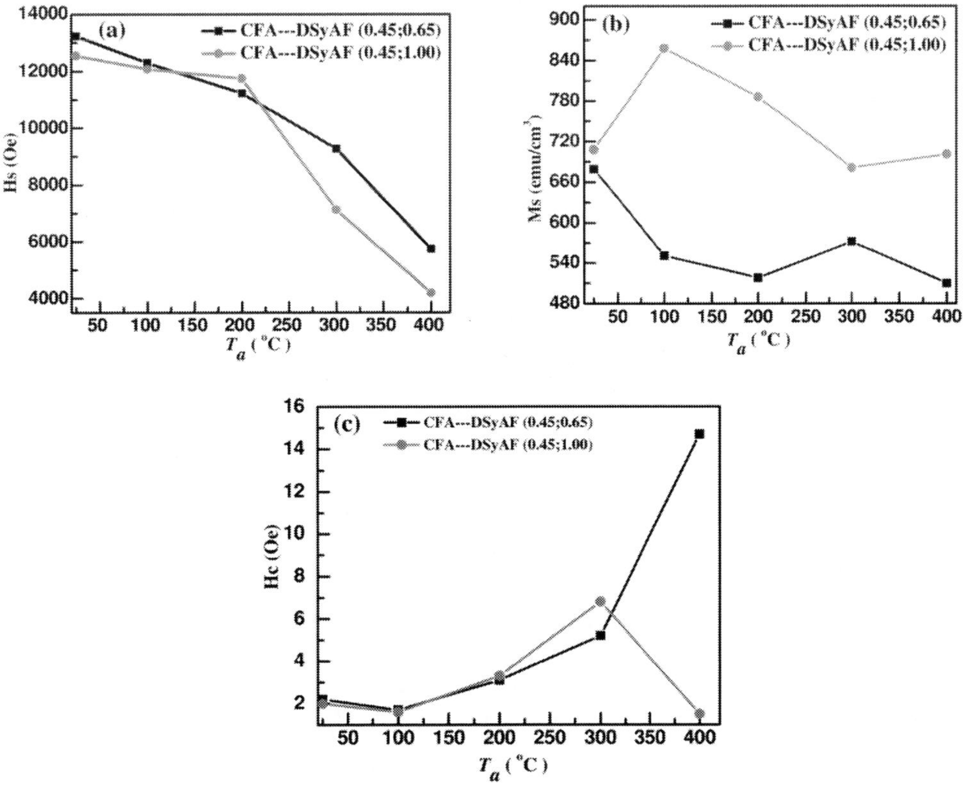

Figure 4. (a) H_s-T_a curves of CFA-DSyAF, (b) M_s-T_a curves of CFA-DSyAF, (c) H_s-T_a curves of CFA-DsyAF.

The CFA-based SyAF (x=0.45) loses antiferromagnetic property after annealing at 200 ℃ [9]. The $Co_{90}Fe_{10}$-based DSyAF (x=0.45, y=1.00) has a low H_s about ~3400 Oe but a high M_s about ~1000 emu/cm^3 after annealing at 300 ℃[10]. Therefore, the CFA-based DSyAFs(x=0.45, y=0.65 and 1.00) have an excellent synthetic antiferromagnetic property and the best thermal stability among all studied SyAFs and DSyAFs candidates, though its antiferromagnetic coupling is weak after 400 ℃ annealing. Accordingly, the CFA-based DSyAF has an enhanced thermal stability up to 400 ℃ and a low M_s comparing with other SyAFs and DSyAFs, which will do favor to ultrahigh storage density. Therefore, the CFA-based DSyAF will have a bright future in spintronic devices due to its excellent thermal stability and the high spin polarized B2-ordered structure of CFA films.

4. Conclusion

We have successfully fabricated the CFA-based DSyAF structures having a stack of CFA/Ru/CFA/Ru/CFA. Compared with all studied SyAFs and DSyAFs candidates, the CFA-based DSyAF has a high saturation magnetic field H_s, a low saturation magnetization M_s, and a low switching field H_{sw}. The enhanced thermal stability up to 400 ℃ of CFA-based DSyAFs allows the phase transformation of CFA to high spin polarized B2-ordered phase.

According to the experimental results, the CFA-based DSyAF is a promising structure for application in spintronics devices.

Acknowledgements

This work was partly supported by the NSFC (Grant Nos. 50701005, 50831002, 50971025), the Key Grant Project (No. 309006) of the Chinese Ministry of Education, the National Basic Research Program of China (Grant No. 2007CB936202), and the Fundamental Research Funds for the Central Universities.

References

[1] Žutić I, Fabian J and Sarma S D 2004 *Rev. Mod. Phys.* **6** 23
[2] Katsnelson M I, Irkhin V Y, Chioncel L, Lichtenstein A I and de Groot R A 2008 *Rev. Mod. Phys.* **80** 315
[3] Galanakis I, Dederichs P H and Papanikolaou N 2002 *Phys. Rev. B* **66** 174429
[4] Kübler J, Fecher G H and Felser C 2007 *Phys. Rev. B* **76** 024414
[5] Inomata K, Ikeda N, Tezuka N, Goto R, Sugimoto S, Wojcik M and Jedryka E 2008 *Sci. Technol. Adv. Mater.* **9** 014101
[6] Zhao Z, Mani P, Mankey G J, Gubbiotti G, Tacchi S, Spizzo F, Lee W T, Yu C T, Pechan M J 2005 *Phys. Rev. B* **71** 104417
[7] Ounadjela K, Muller D, Dinia A, Arbaoui A, Panissod P and Suran G 1992 *Phys. Rev. B* **45** 7768
[8] Fecher G H, Kandpal H C, Wurmehl S, Felser C and Schönhense G 2006 *J. Appl. Phys.* **99** 08J106
[9] Xu X G, Zhang D L, Li X Q, Bao J, Jiang Y and Jalil M B A 2009 *J. Appl. Phys.* **106** 123902
[10] Bao J, Xu X G, Liu Q L, Pan L Q, Yu G H, Jiang Y and Jalil M B A 2008 *J. Appl. Phys.* **103** 07F509

Journal of Physics: Conference Series **263** (2011) 012024 doi:10.1088/1742-6596/263/1/012024

Pulse laser deposition of $BiFeO_3$ films by polished targets

Weitao Su[1,*], Kaixin Song[2], and Dexuan Huo[3]
1, Institute of Material Physics, Hangzhou Dianzi University, 310018 Hangzhou
2, College of Electronics and Information, Hangzhou Dianzi University, 310018 Hangzhou
3, Institute of Material Physics, Hangzhou Dianzi University, 310018 Hangzhou

Email: suweitao@hdu.edu.cn

Abstract. $BiFeO_3$ films had been deposited by pulse laser deposition using $BiFeO_3$ target. Nearly pure phase $BiFeO_3$ film had been grown simply by polishing target before deposition. X-ray diffraction (XRD), UV-visible-near infrared transmission spectrum and magnetization-magnetic field intensity (M-H) hysteresis loops had been measured. It was found that $BiFeO_3$ film deposited at 700° C by polishing target had (010) preferred growth orientation. The bandgap is determined to be 2.25eV. The magnetic domain distribution had been investigated by magnetic force atomic force microcopy.

1. Introduction

Multiferroics, which exhibit both electric and magnetic orders in single phase, provide new opportunity to design multifunctional magneto electronic devices. In recent years, $BiFeO_3$ has been intensively researched for its potential application in multifunctional devices due to the coexistence of ferroelectricity (Curie temperature 830°C) and antiferromagnetisim (Neel temperature 370 °C) [1-3]. As a new lead free ferroelectric material, the polarization value of $BiFeO_3$ film can be as high as 110μc/cm^2, which is close to the polarization value of PbTiZrO3[4].

$BiFeO_3$ film has been successfully deposited by many researchers using sputtering [5, 6], pulse laser ablation [7-9], sol-gel [4, 10], etc. Pure phase and well crystalline film is very essential for multiferroics application. Unfortunately, impurity phases, usually $Bi_2Fe_4O_9$ and Fe_2O_3, often emerge in the films and deteriorate the multiferroic properties. In vacuum deposition, bismuth rich targets are often used to overcome bismuth loss in order to deposit stoichiometric $BiFeO_3$ film [6]. Other growth parameters, such as oxygen pressure [5] and growth rate should also be carefully controlled. In our research, we find pure phase $BiFeO_3$ films could be deposited by polishing target before deposition. This method is much easier and time consuming for deposition of $BiFeO_3$ films.

2. Experimental details

Sintered Stoichiometric $BiFeO_3$ target were used to deposit $BiFeO_3$ films on quartz substrate by pulse laser deposition (excitation wavelength 193nm). Quartz substrates were carefully cleaned by distilled water and then by anhydrous ethanol. The films with thickness about 80nm were deposited at different growth temperature from ambient temperature to 700 °C. In order to control the phase of film, $BiFeO_3$ film was deposited at 700 °C with polished target. $BiFeO_3$ film was also deposited at 700 °C without polishing target. Crystal structure of the films was characterized by XRD from 15° to 70°. The transmission spectra of the films were measured from 320nm to 2500nm.Optical constants (*n* and *k*) were calculated by Lorentz oscillator model. Bandgap of the film were obtained by the cross point of tangential line of plot α^2-$h\gamma$ with energy axis. Magnetic domain distribution was investigated by magnetic force model of atomic force microscopy. Magnetization-magnetic field intensity (M-H) hysteresis loops were measured at low temperature at 298K, 273K 150K and 80K.

3. Results and discussions

As is depicted previously, bismuth loss is inevitable during vacuum deposition of $BiFeO_3$ films. Loss of bismuth always leads to the growth of $Bi_2Fe_4O_9$ and other impurity phase, which had also happened in our experiment. As is shown in figure.1, $Bi_2Fe_4O_9$ phase is predominant for the films deposited at all temperature without polishing the target. However, $BiFeO_3$ becomes the predominant phase if the target is polished before deposition, which is shown as curve E in figure.1. The $BiFeO_3$ phase shows (010) preferred growth orientation. It is thought that the loss of bismuth only occurs at the surface of target. Polishing process could make a stoichiometric fresh surface. Therefore the pulse laser ablation could make a stoichiometric $BiFeO_3$ phase and reduce the growth of impurity f $Bi_2Fe_4O_9$ phase.

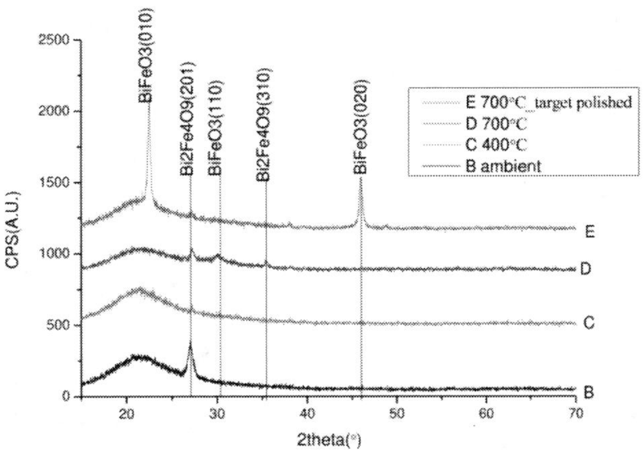

Figure 1. XRD spectrum of different samples deposited at different growth condition.

Optical transmission property in visible light range is very important for its potential application in magnetic-optic storage. In this experiment, UV-visible-near infrared transmission spectra of the samples have been measured and shown in figure.2 It is found the film deposited at different condition have good transmission from about 700nm to the near infrared. It is also found that $BiFeO_3$ films deposited at the center and rim of the glass substrate have different transmission propertiés, which indicates non-uniform of film thickness and even the different absorption. It is reasonable if the non-uniform intrinsic property of pulse laser ablation is considered. The cutoff wavelength of $BiFeO_3$ film at the center ($BiFeO_3$_700center in fig.2) and rim ($BiFeO_3$_700rim in fig.2) of substrate does not change very much, which indicates they have nearly the same band gap. It is much interesting that Bi2Fe4O9 phase have larger band gap than $BiFeO_3$, which is also shown in figure.2

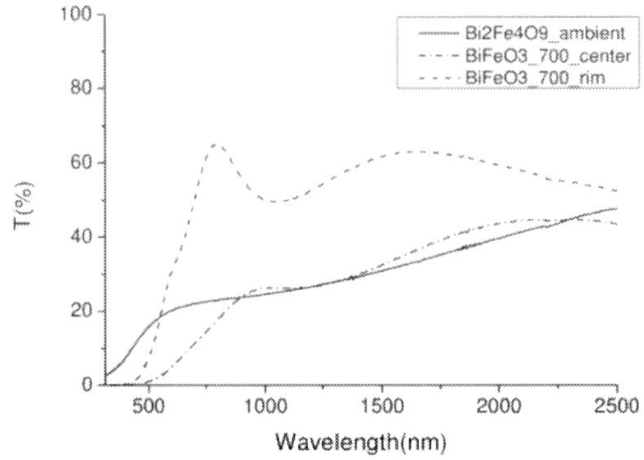

Figure 2. UV-visible-near infrared transmission spectrum of BiFeO3 films

The optical constants of $BiFeO_3$ films are also calculated by fitting the transmission spectra ($BiFeO_3$_700rim curve in fig.2) using Lorentz oscillator model. The refractive index (n) and extinction coefficient (k) is about 2.6 and 0.32 at 635nm, respectively, which is shown in figure.3. The absorption coefficient can be calculated by $\alpha=4\pi k/\lambda$, where λ is the wavelength. The optical band gap can be obtained by intersect of tangential line at energy axis in the plot α^2-hv, where hv is optical energy in eV. The optical band gap of $BiFeO_3$ films is about 2.25eV, which is shown in figure.4.

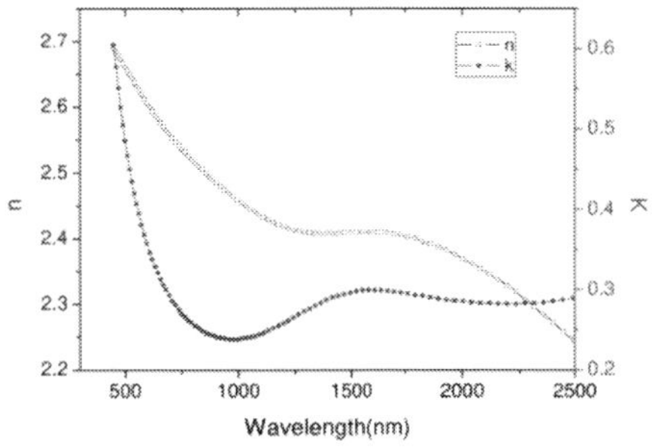

Figure 3. Calculated optical constants of $BiFeO_3$ films

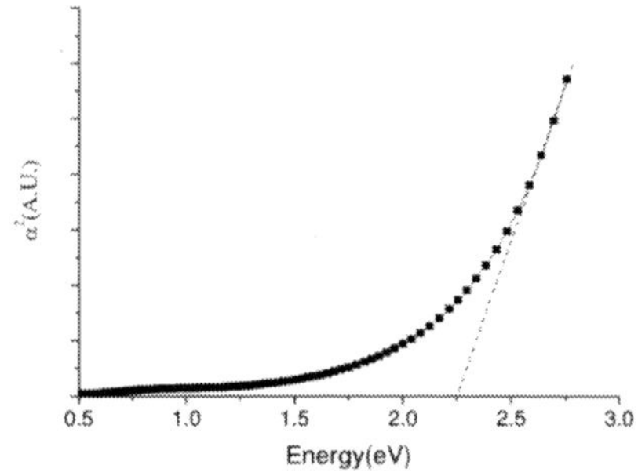

Figure 4. Band gap plot of $BiFeO_3$ films

Magnetic force mode of atomic force microscopy was used to investigate magnetic domain distribution of $BiFeO_3$ films. In first step, the top morphology image was recorded by constant-current mode, which is shown in figure.5. The surface is quite smooth and covered with homogeneous crystal particles. The crystal particle size of $BiFeO_3$ film is about 200nm. In second step, the tip was lift 80nm higher than the surface of film. Far field force image was recorded as magnetic force image, which is shown in figure.6. The size of irregular magnetic domain is estimated to be about 400-500nm.

Figure 5 AFM top morphology picture of $BiFeO_3$ films deposited at 700°C with polished target.

Figure 6. MFM morphology picture of $BiFeO_3$ films deposited at 700°C with polished target.

Magnetization (M) vs. magnetic field intensity hysteresis loops of $BiFeO_3$ films at different temperature are shown in figure.7. In this experiment, the magnetization of quartz substrate could not be neglected because magnetization of $BiFeO_3$ is very weak for its thin film thickness. Magnetization of $BiFeO_3$ film overlaps with quartz substrate, which leads to the rotation of magnetic hysteresis loops. The area of hysteresis loop increases with decreasing of measurement temperature. The rotation of magnetic hysteresis loop makes the determination of saturated (remnant) magnetization much difficult. However, one can find that the saturated magnetization increase qualitatively with decrease of temperature.

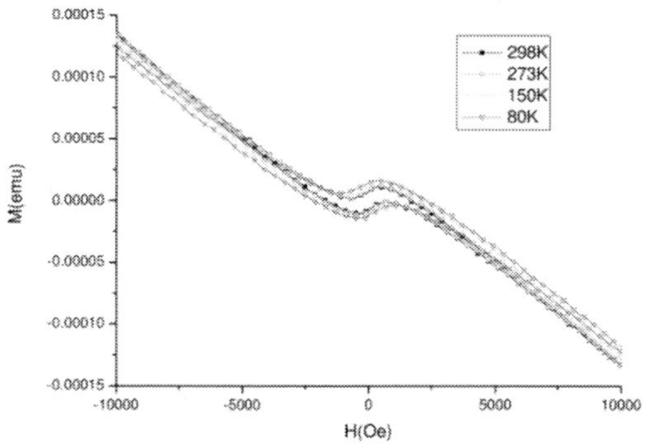

Figure 7. Magnetization hysteresis loops of $BiFeO_3$ film on quartz substrate at different temperature.

4. Conclusions

$BiFeO_3$ films had been deposited on quartz substrate at different deposition temperature. It was found that nearly pure phase $BiFeO_3$ films could be easily deposited by polishing the targets before deposition. The films deposited at 700 °C by polishing targets have (010) growth orientation. The transmission spectrum were measured and fitted to get absorption coefficients. The bandgap of $BiFeO_3$ film was calculated to be 2.25eV. The magnetic domain distribution was measured by magnetic force microscopy. It was found the saturated magnetization of $BiFeO_3$ film was increased at low temperature.

Acknowledgements

The authors gratefully acknowledge financial support from Zhejiang Provincial Natural Science Foundation of China (Grant No. Y1090146)

References

[1] Khomchenko V A, Kiselev D A, Kopcewicz M, Maglione M, Shvartsman V V, Borisov P, Kleemann W, Lopes A M L, G.Pogorelov Y, Araujo J P, Rubinger R M, Sobolev N A, Vieira J M and Kholkin A L 2009 *J. Magn. Magn. Mater.* **321** 1692

[2] Wen Z, Hu G, Fan S, Yang C, Wu W, Zhou Y, Chen X and Cui S 2009 *Thin Solid Films* **517** 4497

[3] Prashanthi K, Chalke B A, Barick K C, Dasc A, Dhiman I and Palkar V R 2009 *Solid State Commun.* **149** 188

[4] Wang Y and Nan C 2009 *Thin Solid Film* **517** 4484

[5] Ternon C, Thery J , Baron T, Ducros C, Sanchette F and Kreisel J 2006 *Thin Solid Films* **515** 481

[6] Qi X, Tsai P, Chen Y, Lin Q, Huang J, Chang W and Chen I 2009 *Thin Solid Films* **517** 5862

[7] Zhang G, Cheng J, Chen R,Yu S and Meng Z 2006 *Trans. Nonferrous Met. Soc. China* **16** s123

[8] Singh P, Park Y A, Sung K, Hur N, Jung J H , Noh W S, Kim J Y, Yoon J and Jo Y, *Solid State Commun.*, article in press

[9] Toupet H, LeMarrec F, Holc J , Kosec M , Vilarhino P and G.Karkut M 2009 *J. Magn. Magn. Mater.* **321** 1702

[10] Liu H, Liu T and Wang X 2009 *Solid State Commun.* **149** 1958

AUTHOR INDEX